STRANGER THAN FICTION:

THE TRUE TIME TRAVEL ADVENTURES
OF
STEVEN L. GIBBS
THE RAINMAN OF TIME TRAVEL

BY

PATRICIA GRIFFIN RESS

ISBN: 0-7596-7707-7

This book is printed on acid free paper.

1stBooks - rev. 11/09/01

INTRODUCTION

This book is the most fantastic thing I have ever written! Fantastic because of the subject matter itself. Fantastic because of the mesmerizing personality of Steven Gibbs and fantastic because of the many things he has encountered in the past 20 years! If it had not been for Steve, I would have remained more than a little hesitant to reveal my own "time slippage" experiences and those of other people close to me. I have come to the conclusion that Steve is somehow channeling his information but from what or from whom, I have no idea! It is more complicated than what many people study in college and even graduate college! In most matters, Steve is an average Nebraska farmer, but in his own narrow realm of wisdom; he is an Edison or Galileo.

A few years ago when I was trying to explain his unique ability to access and to understand knowledge way beyond his level of education, I prefaced my explanation with a question. I asked, "did you ever see the movie *Rainman*?" The person to whom I was speaking immediately replied, "so Gibbs is the *Rainman* of time travel." Although it wasn't exactly what I intended to say, it **did** fit and so I agreed and the description stuck.

This book is in two basic parts; the first part explains how and why I first contacted Steve, some of our joint experiences with mutual friends, by ourselves, and with each other and concerns itself with other time travel/slippage accounts I have heard about. The second part includes some magazine articles I have written about Steve for publications such as Prediction Magazine and Uri Geller's Encounters Magazine.

It also includes letters he has written to me or received from people who have used his time travel machines with successful results! Lastly, there are excerpts from a few of his own reports, which he sells individually through catalogs.

Steve has a web page and advertisements that run monthly in several magazines which draw inquires from all over the world. He has also been a guest several times on A.M. Coast-to-Coast talk show with legend Art Bell, and then with Mike Siegal. He has even attracted the attention of Canadian and South American television stations and wowed the audience of the Mike Jarmus Show. For many years this show had the largest audience of short-wave listeners in the New York City area. Steve has been contacted by at least a dozen radio stations and has given numerous radio interviews, and has been the subject of many articles in various publications as well.

Mulder and Scully (of the X-Files) would say that Gibbs' fans "want to believe", but why? Surely most of us who would want to escape the "hum-drum" of our everyday lives realize that time travel to bygone eras or to a future of unknown quantities cannot actually go, just yet! Maybe for the military, or the word's ruling elite, but Steve Gibbs has shown us that the average person **can** time travel too, and he has shown us several ways to do so!

What lies and half-truths have the multinationals and the politicians hidden from the masses? If lies could be uncovered and revealed, could we be freed from our bondage? If time travel is something we can **all** access then couldn't we all see the truth? That would empower each and every one of us!

This is precisely why Steve Gibbs has attracted so much attention and admiration. He is strange, but he is also "everyman", he is one of us!

This is the story of my friend, Steve Gibbs, Nebraska's time traveler extraordinaire! Steve has written several books, attracted the attention of academics and curiosity seekers alike, and has been a guest on numerous talk shows throughout the United States, most notably Art Bell's Coast-to-Coast. To some he seems to be the quintessential "*country bumpkin.*" To others he is a hermit savant who has somehow accessed knowledge beyond us all. No matter what you might think of him personally, he is indeed, a fascinating subject.

This book has nothing to do with politics, please keep this in mind no matter what stripe your own persuasion. But it did **begin** with politics and I will tell you how.

I live in Nebraska, but I am not from here, or thank God, of here! People here will never admit it, but they all belong to a peculiar religion called the Republican Party. There is nothing wrong with the Republican Party; some of my best friends are Republicans. But in Nebraska you see weird intellectual gymnastics to persuade you that you cannot be Christian, God-fearing, well educated, tax paying or generally worthwhile, unless you **are** Republican. Case in point. I once heard a Nebraska farmer say that "Jesus never said give all you have to the poor and follow me." He insisted that such was a "dirty, liberal, commie idea." Oh really? I wonder what Bible gave him that impression or information! I heard another Nebraska businessman say that the "Bible said something about the poor are always with you."

Therefore, he concluded, "the Republicans were quite right about not wanting to help the poor for fear it was going against the will of God!" Whew! What a conclusion! I thought that when Jesus remarked that the poor were always around, that he did so in exasperation as if he wished that someone would do something, yet knew they would not! Well, that's how little I know.

Well, at least Huskers are consistently illogical. For example, I could never understand why someone who claimed to be "pro-life" could turn right around and oppose universal access to healthcare. Don't babies need to be protected from diseases too?

Now, as a writer who is forced to endure this constant right-wing assault, it has crossed my mind that it would be an interesting story if someone could actually contact the "great beyond" and hear from the "horse's mouth," so to speak, what "true" Christianity is as opposed to what Nebraska's version of it, actually is.

In the early 1990's I wrote for The Midlands Business Journal and Omaha Metro Update. While doing this, I met Aage (pronounced Auggie) Nost, a pilot, talk show co-host and publisher of The Constitutional Liberator. Through him, I learned a lot about UFO secrecy, government conspiracy and other topics no one else wants to "touch with a ten-foot pole!" I told Aage about my story plans and he told me about a patent he had seen. A patent held by Thomas Edison for a machine that could contact the dead!

While this sounded intriguing, I never could find this mysterious patent, or device, but instead discovered the experiences of a European researcher who dealt with EVP (electronic voice phenomena). This, you might recognize, is what is left on blank tapes that are allowed to run undisturbed in certain locations such as haunted houses, graveyards, etc. But EVP is not message specific. You cannot pose a deep theological question (or any other kind of question) and get a complex, detailed, and definite answer. I mentally put EVP in my "possibility" drawer and looked for something else. That's when Aage suggested that I contact Steve Gibbs.

I had no idea what to expect from Steve. Our few telephone conversations were cordial, informational, and pleasant, but I could not see how this farmer/electrician with little formal education could possibly construct a machine with healing abilities, the ability to access the spirit world and other dimensions and travels through time...that last one was mind-boggling! However, it turned out to be more important than the others are!

Once I met Steve, I discovered that my original idea, though perhaps interesting, had none of the potential to get to the "*truth of the universe*" as did Steve's primary invention: **The Hyper-Dimensional Resonator.** Why worry about what anonymous voices on tapes might say about contemporary religion, or lack thereof! Here was the ability to access the light, to gain answers and insights that were above the questions of whether or not my contemporaries were practicing what they claimed to believe. I decided to let them live in their self-paralyzing delusions giving over any individual thought to their local power brokers, and clergymen, and oh, yes, the local ward boss of the ever omnipotent Republican Party!

BUT WHERE DOES HE GET HIS INFORMATION?

By the time I got to meet Steve, I was well into writing my second UFO book (along with a romance under a Nome de plume). When I told some of my coworkers that I was going to interview Steve for an article in Prediction Magazine they were very interested in the subject of time-travel. "Where did he study? Where did he get his degree?" These were frequently asked questions. You can imagine the "flack" I took when I revealed that Steve had only an average education with a correspondence course in electronics! I was laughed at for being so foolish as to believe him.

When I met Steve for the interview, I began to regret the entire thing! Maybe Aage had heard wrong. How in the world would this simple Nebraska farm boy know anything about time travel? But he did! Though Steve did not display more than an ordinary knowledge of most subjects, when it came to physics and quantum mechanics, I had the feeling I was in the presence of a true scientist/cosmologist! He totally understood their concepts and demonstrated an ability to, as Stephen Hawking once explained, *"to see the physical laws he was trying to explain"*. From what Steve told me, the information comes to him almost as though it is through a channeler. "*The Lord gives me this information and I take it and use it to help others,"* he explained.

STRANGE THINGS START HAPPENING

So I was impressed with what Steve knew, but could he deliver? Could he prove to me that he had actually traveled through time and/or other dimensions? I found out soon enough!

My article about Steve attracted a lot of interest and attention. My friend, Aage, came to visit the next time Steve was in Omaha. He brought his wife, Barbara with him. She had been suffering from an eye ailment and failing vision. He wanted to test the healing function of Steve's **Hyper-Dimensional Resonator.** Although it was a one-time shot, it did seem to help her for a while. Aage also tried the resonator but did not wish to travel physically through time. He claimed he traveled ahead in consciousness where he saw the Pope in failing health, Frank Sinatra dying, the United States involved in Kosovo, and the ethics problems for Bill Clinton. All of these, in time, came true and were not indications of anything more than a consciousness expanding function which Steve had already known about.

What happened next threw us all for a loop! Steve had gone to his motel for the night, Barb and Aage had gone home, my son and husband had gone to bed. It was 2:00 a.m. and I went to the kitchen for a glass of water. I looked out across our parking lot and saw three clouds shaped perfectly like flying saucers! Normally I would not have paid that much attention to such forms, except an abductee I knew had recently told me that UFO's often take the forms of clouds as camouflage. Suddenly frightened, I turned on the light switch and saw sparks dance across my kitchen light fixture!

It had never happened before, so I screamed! My husband and son were in the kitchen within seconds! When I told them to look across the parking lot, they asked me what for, so I looked again and the UFO shaped clouds were gone! Now clouds normally dissipate, I realize that, but not in a few seconds! Later Steve explained that some aliens probably had seen the resonator working and had come down to investigate it. I tend to think he was right!

Shortly after that my granddaughter came to visit. She went to bed to take a nap and when I went in to check on her, she was gone! However, when I came back a few minutes later she was there! Steve said I had simply shifted universes! My granddaughter insisted that she had been napping all the time and hadn't gone anywhere! I had not left the apartment!

Steve seems to be able to call down cosmic forces, which Heaven supplies to protect him or give him warnings. After Steve first appeared on the Art Bell Coast-to-Coast Show, he was contacted by a talent scout, the son of a prominent national celebrity. Steve agreed to meet with the man and introduced me to him as his "agent". The man seemed to have more of an interest in Steve's technology than in him as phenomenon and Steve was very suspicious that he might be from the government trying to get information on how much Steve knew. Steve refused to meet him again or sign any contracts because of a rune reading he was given telling him to beware of this man. At the time I thought he was being paranoid, then I later found out that the man's father did, indeed, have governmental connections. I never distrusted Steve again!

Then there was the matter of a lady named "Charley" who seemed very interested in Steve's time-travel experience.

While Steve talked to her on the phone, he declined to meet her because he "sensed" that she could be a mole for a secret agency. Again I felt exasperated! Steve was being silly! She wasn't the kind of person to be involved in such things. Then she inadvertently told me one day that her husband had once worked in military intelligence and with the secret service. I had never guessed. Steve was right on target again!

On yet another occasion, Steve told of going into the future while in one of his older cars. He wound up in a nearby town by an unfamiliar building. When he went there 2 weeks later, the building had not been built yet, but when he went there 3 years later it had!

One of the last things I want to reveal in this section includes Steve's contact with a fellow time-traveler. This person sounded to me like someone out of

World War II intelligence. When Steve told me his name, I nearly flipped! It was the name of a famous figure from that era!

On a recent Art Bell Show, Art read a fax from a young man who had tried Steve's time travel machine and had seen himself go several minutes into the future! This same young man was later so harassed that he had to move to a foreign country for fear of his life.

With Steve, we have someone who seems to know about things happening in the future or from before he was born. We also have the people who seem overly concerned about his knowledge and how much he **DOES** know. If he is not "onto something" why the fuss? If his ideas are so crazy, why the bother? Bad ideas run their course, or are harmless and produce nothing. But someone, somewhere knows that Steve Gibbs is onto something, and that made him an exciting person to investigate further!

I REMEMBER SOME UNSETTLING EXPERIENCES FROM THE PAST

After my initial experiences with Steve, things from my past suddenly began to come flooding back to me as though he had tapped a reservoir of hidden memories. The first thing I recalled was a strange incident, which had happened to me in 1986 when I was writing for a countywide newspaper in the South.

One of my many assignments was to report on the town council meetings in the various communities served by my employer. Sometimes I was faced with the dilemma of being held responsible for the news material from 2 or more council meetings occurring on the same night! At those times, I would attend the meeting that was most controversial or affected the largest number of people. Then I would stop by the town halls where the other meetings had been and get the notes the town secretaries had taken, so I could do a report from them.

On the evening in question, I had just completed the first meeting and stopped to collect the notes from the others. It was 11:00 p.m. and I was finally headed back to the newspaper office in the main county seat town where I worked. As I passed the highway sign with arrows pointing straight ahead and indicating the number of miles I had left, I drew a sigh of relief! Had I been in an airplane, I could have set the automatic pilot for straight ahead and forgotten everything. As I drove along, I listened to the sounds of the night, the low music from the radio and the light grind of tires on the pavement.

Then suddenly, it was almost totally silent! The van in which I was travelling seemed to be floating in an odd colored mist. Ahead of me I saw nothing, but to my right I saw an elderly black lady chopping wood on an aging tree stump. Her face was stern, almost foreboding, and was illuminated by a large kerosene lamp on the ground. Nearby her stood a harnessed mule and when she looked up she stared through me as if she hadn't seen me at all! Behind her was an old brown shack, older and more dilapidated than anything I had previously seen in the area. At this point, I would like to insert a bit of information about the rural south today. There are **many** poor southerners, both black and white. But they **do** live in the 20th (now 21st) century. They have trucks and tractors albeit old ones and they have electricity. They may only have it on one or two rooms, but they **do** have it. Their clothing sometimes comes from catalogs or thrift stores, but it is also contemporary. But the scene that had startled me was circa 1860, and the black lady I had seen was dressed like the beloved "Mammie" from Gone with the Wind.

In what seemed like only a few minutes, I was near the center of town where my papers' office was located. My mileage indicated that I had driven the same distance as usual, so I chalked up my experience to exhaustion, turned in my film and stories, and went home for some sleep. A few days later I had occasion to drive that same highway again, only in the daylight hours. Although I was absolutely certain I hadn't left the main road (or the odometer would have indicated so), I drove around the immediate area, but saw nothing that even faintly resembled what I had seen that mysterious evening I didn't think much about it again, until I met Steve Gibbs. He told me that if I had gotten out of the van, I would have locked myself into that time period! The van would have been kinds of a time machine.

During the winter of 1997, I attended a couple of parties at the home of an artist friend in Omaha. It had been colder and snowier than usual, so we had gotten together to watch some interesting videos and talk. Usually, people would pair off, according to their interests, and on one occasion I found myself paired off with a young couple who had some very strange experiences in our own city.

The young lady to whom I spoke seemed more excited about her experiences than did her more quiet husband who listened with a kind of polite resignation as she related two experiences she had found most perplexing. On one occasion she had been in the historic **Old Market** area of downtown Omaha, Nebraska. She had seen apparitions of people clothed in garb from the end of the 19th century. They seemed to fade in and out with the surroundings. Unable to find a logical explanation for such things, she chalked it up to being exhausted or somehow misinterpreting what she had seen. But on another occasion in the same general area, she was driving her car when she happened to glance into the rear view mirror and saw cars, which were models from the 1940's and 1950's! She was so startled by this that she turned around and looked again only to see more contemporary units whizzing by her on the street! When I told Steve about this he suggested that something might have caused her to shift universes momentarily. I was shocked that one could shift dimensions or "time frames" as easily as they could snap a picture. But Steve explained that he did, indeed, do this and nearly on a daily basis!

But before we get into any of that, let me tell you a bizarre story that my artist friend claimed happened to her in September of 2000 AD.

Since she had been in Pennsylvania for most of the summer, my friend was shocked when she received an electrical bill that was not only much higher than it had ever been, but also higher for the time she had been in away! Well utility companies **do** make mistakes, so although they would not back down on the amount owed, they sent out a repairman to correct any problems which could be giving her an "inappropriately large" bill. The repairman arrived a few days later in a company truck with the proper identification in hand! After replacing the meter, he informed my friend that it had been over 30 years old and in poor working condition. Before he walked out her front door, he assured her that everything would be just fine, but it wasn't!

When she received her October bill, it was again out of sight! By then she was totally disgusted with the utility company! When she called them again, they agreed that she had just had a new meter installed, but not in the previous month, they claimed it was the previous **year**. When she gave the name of the repairman who replaced it, they insisted that no one by that name had been working for the company in recent time! My friend called and asked me, "Pat, am I in the twilight zone or what?" My response was "or what." Then I told her about some strange time anomalies that had seemed to exist only 40 miles away in Lincoln, Nebraska.

In 1991 my husband and I, along with our youngest son Eric, had been living in Omaha for about 2 years. We had moved there from Lincoln, Nebraska where the job market had been abysmal.

My son had a friend whose aunt was an English professor there and had managed to get a better teaching position in Omaha. He and Eric were going to spend the weekend helping the aunt move to Omaha with the help of his grandparents. My husband and I said it would be all right with us and gave the matter no more thought. But when Eric came home at the end of the weekend, he had a very strange story to tell us about what had happened to them all in a part of Lincoln known as "University Place." Apparently they had stopped at a stop sign and a large semi-truck had pulled up next to them. It had seemed to come out of nowhere. Eric asked his friend a question and had turned away for only a few seconds. When he looked back, the semi was gone! Could a huge truck like that have taken off at such speed and gone such a distance that no one in the car could see where it had gone? According to the grandfather, **no way possible!** So was it a truck? Had it slipped in or out of another universe? No one seems to know, but it wasn't the first time that a time travel tale had come out of University Place.

I remembered an even stranger story from a few years before! In 1977-1978 I worked for the Sun Newspapers in Lincoln as a feature writer/photographer. As Halloween approached, I set out to find some spooky stories I felt the public would enjoy! One of the other writers in the newsroom asked me if I had ever heard of the "Ghost of Nebraska Wesleyan College?" She suggested that it might make a good story for the Halloween edition, so I called the college and they referred me to a professor who was still on the faculty and who had been teaching at NWU at the time of the "incident".

To my 32-year-old eyes, Dr. David Mickey was "old." But for the story I was doing, "old" was something I was looking for and his graying hair only added a kind of distinction to his demeanor. Before we got to the ghost story, we talked about Charles Starkweather, the spree killer who had terrorized Nebraska two decades earlier and had since become the topic of several books and television programs. Dr. Mickey remembered details of that time that most of us had forgotten. His mind was as sharp as a steel trap, so when he told me about the ghost story, I knew I was hearing it from a very capable and credible witness.

Without going into all the details, here is what basically happened. In the last 1950's, a secretary for the music department at NWU began what she believed would be an ordinary day. But when she opened the door to the library, she was greeted with a strange sensation. Everything seemed different. As she looked over towards the shelves, she saw an older woman with her hair done up in a bun and wearing clothing from a different time period. Puzzled by all this, the secretary went over to her desk, but when she glanced out the window behind it, the entire landscape had changed! At that point, she was sufficiently traumatized and went running from the room! Later on, after calming down and telling her story to some other observers, she was shown yearbooks from 20 and 30 years earlier. The landscapes of the campus in them were exactly as she had seen them earlier that morning and she recognized the old woman she had seen! She had been a music professor who had died in the office across the hall from the library at exactly 9:00 a.m. nearly 30 years ago! The incident was so bizarre that it attracted the attention of several prominent psychological clinics, but to this day it has never been fully explained to everyone's satisfaction.

After this story ran, several people from the same area of Lincoln who had experienced similar ghostly or time slippage experiences contacted me. Now, after all these years, I am beginning to believe that the entire State of Nebraska is situated on some kind of time portal where people, ghosts, events and even places slip in and out of reality.

Indeed, even Donna Butts, the famous abductee / contactee of the UFO Butts-Corder case, said that she had been told by a group of aliens calling themselves the Americans that a "transmutational channel" had opened over Nebraska and Kansas. Perhaps it is no coincidence that Steve Gibbs is from Nebraska and so is author Kathleen Keating, who sees the Biblical rapture occurring in a manner not inconsistent with the workings of Steve Gibbs time machine.

In her international bestseller, The Final Warning, Keating discusses how the rapture could possibly happen. Unlike fundamentalist Christians who believe the faithful will all be zapped up into Heaven at once, Keating believes the rapture could happen in increments, a few here, a few there, etc. If she is right, why couldn't people be taken into other dimensions so they will not suffer the horrors that are slated to begin on earth?

Keating also brings up another time-related subject something we have all been sensing, the possibility that time is, in fact, speeding up. She points out that in the end, time may be speeding up so that good people will not have to suffer as much. (An Australian woman who claimed visitations from the Virgin Mary was told the same thing).

Keating points out a phrase in the Bible which clearly states that if the days (in the end times) had not been cut short, even the very elect would be lost! After all, if we are all gone who will be here for Jesus to rule during that wonderful time referred to as "the reign?" (It is interesting to note that many UFO abductees / contactees claim that aliens have the ability to either shorten or prolong time or even freeze it!)

Years ago when I was a graduate student at the University of Iowa in Iowa City, I had a friend who was always insisting that she and I were both "out of sync" with the time dimension. One example of this happened when I placed a roast in the oven and set a timer for 90 minutes. My friend called me and we talked for over 2 hours. She related each time that a new television program came over her set. After awhile, certainly well after the 90 minutes, I began to worry that my timer didn't work, so I excused myself and checked. It was still ticking and the roast will still cooking. After what would have logically been three hours, the timer rang, I took the roast from the oven and everything was fine. I tested my timer later that same evening and again the next day and it was right to the second in accuracy! Is time just a sequence of events that happen routinely in this dimension? Or is it a period of an activity going on which only **seems** like it takes a long or short time depending upon how we feel about it? If we **do** "pop in and out" of other dimensions, does time flow differently in those dimensions? Or as Steve claims, is time something that exists **only** in **this** dimension?

If there is any aspect of time, which is personality specific, it is the aspect of astrology or the exact time and place of one's birth and how it affects destiny.

Scientists like the late Dr. Carl Sagan have made fun of astrology because they find it difficult to believe that the planets have any effect over one's life. Dr. Sagan once said, "the doctor who delivered you has more of an influence on your life than say, Jupiter, etc." Not so fast, Dr. Sagan! The gravitational pull of Pluto may not have any direct effect, **but** the timing of things is crucial! In

Recent years a team of French scientists made a long-term scientific study of birth times and the life pathways of several diverse people and guess what? **When** you were born (and where in terms of latitude and longitude) actually **does** affect you in almost every way; health wise, career wise, personality wise, etc. One of my friends told me, "Pat you are never going to believe astrology, or understand it, until you've met your astral twin." I asked her what an "astral twin" was and she explained that it would be someone born the same day, month, year and hour as me. So where would I find such a person?

In 1968, I took about 10 days off from my study schedule to visit one of my favorite places Quebec, Canada. On my way home, I ran into a young man who had boarded the bus in Montreal. He and I had an interesting conversation and somehow discovered that we had both been born the same day, month, year and (when adjusted for differences) the same hour!

We talked until dawn when we arrived in Chicago, and had found we were so similar it was frightening! Our fathers had the same occupations. We were both interested in the same things academically, the same life events had happened to us at the same ages! Then here comes the clincher! We both got off the bus, said goodbye to each other and began walking away. I put my right hand in my pocket to make sure I still had my train ticket for home and discovered an extra dollar, the exact amount I had borrowed from him in Detroit where we had eaten a quick lunch. I turned to tell him I could pay him back, but he was nowhere to be seen! Like the semi truck in Lincoln, he had been walking in a direction where I should have seen him no matter where he would have gone, but he was **nowhere!!!** He seemed to evaporate into thin air! I have often wondered if he had been a messenger or (possibly) an angel from another dimension verifying the relationship we all have to one another via time. Who knows?

Later on in this book you will read what Steve has to say about finding your "other self" and how this entity can destroy you if it comes into direct contact with you. Could this mean that there is a "lock" on one's soul for a particular time frame within a particular dimension? Steve seems to think it does and that one should not time travel unless they are "destined" to do so. They could be "shorted out" by someone from a different place who would probably have a different electro-magnetic polarity. Then, if they do have this different "polarity" (for lack of a better word), can they survive in this dimension and this time frame for short amounts of time?

Recently, Steve revealed to me that during a trip to a nearby small city he noticed 2 men who appeared to be classic M.I.B.'s (Men in Black). But no one else seemed to be able to see them! Was this because they did not "belong" in this dimension? If not, where did they come from? And why?

In fact, right now, at this point in history, a lot of strange things seem to phase in and out of this dimension. One clear example is the Chupacabra or "goat sucker" which manages to kill and mutilate livestock yet disappears without a trace!

When I first met Steve, he told me that when he had entered another dimension, he had found both the Mayans and the Incas! Is this crazy? Before you decide, read on!

According to Dr. Lianna Carbon (September issue of Emerging Awareness Journal) the Q'ero are the last of the Incas, a tribe of 600, who sought refuge at altitudes of 14,000 to escape the Spanish conquistadors. The Q'ero elders have preserved a sacred prophecy of a great change in which the world wold be turned right side up and harmony and order would be restored.

The Q'eros have rights of Mosoq Karpay which represents the end of one's relationship in time. I find this very interesting since Steve has frequently reminded me that time exists only in this dimension! Why it **does** and why it will no longer have a relationship with us are major questions in and of themselves. But the Q'ero further believe that the doorways between the worlds are opening again, holes in time that we can step through and beyond!

They believe that there will be a tear in the fabric of time itself that will let us describe ourselves in terms of what we are becoming. Is this because by then we will realize that time/dimensional travel has allowed us to accurately see or experience the future?

On November 1, 2000, a guest on AM Coast to Coast with Mike Siegal revealed that a group of scientists had discovered a way to enter another dimension and had actually gone there twice before deciding to go there permanently a third and final time! These were serious intellectuals who were fed up with the dictatorship of the military industrial complex. The guest discussed the method this group had allegedly used. It was a combination of machine and mindset working together! It is a method very similar to the **Hyper-Dimensional Resonator** developed by Steve Gibbs!

Although the guest insisted that time travel and inter-dimensional travel are two different issues, Steve believes they are linked. In fact, he believes that when **he** time travels, he bounces back and forth between the universes on the immediate right and left of this one.

Back in 1987, "new agers" celebrated the so called "Harmonic Convergence" and since then I have heard several metaphysical sources claim that mankind is "ascending" to a higher level of enlightenment in which those of us who are truly on a righteous path will vibrate at higher and higher levels where we have only to think something and it appears, available for us! The wicked and evil among us who crave money, power, and self-elevation as opposed to goodness, kindness and justice, will get to stay here among the lower vibrations to fight for their winnings. They believe whoever dies with the most toys, wins (as the bumper sticker implies)! The rest of us will no longer play that little game!

Are people and entities from other dimensions as eager to help us out or "enlighten" **us,** as we are to find them? (Or are **we** eager to find them?)

Well, as I mentioned before, they do occasionally wind up in our universe for reasons not always apparent to us. For example, let me tell you about a book that was sent to me by...?

In one of many catalogs I receive on a regular basis, I saw a review of The Time Travel Handbook by David Hatcher Childress. (Incidentally, I highly recommend this book for those readers who want a real *"nuts and bolts"* scientific approach to time travel, it is definitely the best around). About 3 weeks later the book appeared in my mailbox sent by a company that carries the book in the East. I immediately called Steve and thanked him for the book, but was surprised to hear him say, "but I never sent that book to you! I was thinking about it for a birthday present for you but that's not until August!" (This was early in the year so my birthday was something of a wait.) Anyhow, I asked my husband, my son, my older daughter and anyone else who may have had the book sent to me, but no one had done so, yet there it was for me to read! Was I going crazy? Perhaps, does someone from somewhere feel I need to know what is in it? Only time will tell!

Part of Steve's **Hyper-Dimensional Resonator** includes a "witness well" which has many purposes. If I'm not feeling up to par, Steve will take a small photograph of me and place it in the well to project white light power to me. The result? He has helped me feel better, perhaps even recover from and get though a plethora of maladies that have affected me over the years since we first met in 1991. But something else happens to me when I am "zapped", my psychic powers kick in. Not in everyday life, but on the astral plane where I visit after falling asleep each night.

Ever since I was a small child, I have often dreamed things that have come true. In fact, I often dream something only to see and hear the exact scripting the next day in real life. My most recent experiences began with the death of a friend of mine nearly 5 years ago. Suffering from a chronic illness, she had made numerous trips to the hospital in recent years and survived them all! On this spring day, however, things were very different. It took me almost 2 days to even find the hospital where she had been taken and even then, it was her cousin who called me and told me what was going on. Anyhow, I had gone to bed on a Thursday night and after my dreams had come and gone, my friend appeared at the foot of my bed. She was dressed in a white nightgown and looked just like she had when we were in high school. She told me she was leaving and that she'd see me again someday and not to worry about her, then she walked away. I was just starting to wake up when the phone rang. It was her cousin calling to tell me she had died. I told her cousin that I knew that my friend had died and told her about the dream. She seemed surprised and probably thinks to this day that I am a "nut case" of some kind. But starting then I began to dream the dreams I had once dreamed so frequently.

Many older people in Omaha remember a funeral home, which served the community for many years; it was called Fitch and Cole. E. Quay Fitch was a member of the family who had owned that mortuary and he was a friend of ours until his death in 1998. Quay, as we called him, was interested in many of the same things that my friends and I enjoyed. After he died, all of us missed him a lot, but he came back to help some of us in his own strange way. Two friends of ours had a washing machine that had given them headaches on a regular basis. Quay had suggested that they try this or that, but they never did. But after he died, my friends swore that they heard him working on their washing machine and the next time they used it, it worked perfectly! Coincidence? Anyhow, after Quay died Steve had suggested that if I wanted knowledge about something to simply ask God for it. So I rather casually (and only half seriously) asked God for greater psychic insight into what was going on in the world. I began to have prophetic dreams where I was shown what would be happening on our physical plane of existence. The alternate universe I frequently visited on the astral plan (or from the astral plan depending on your particular view) is like the place in the movie **dark city**. Everyone is caught in a kind of existential rut where nothing begins or ends and people are only assigned different roles by transferring memories. But in this alternate universe where I have no reference points, I have the certainty that I am in Kansas City (or at least the counterpart of Kansas City) and I am living next door to Quay Fitch. In this universe, some of my worst fears are realized. In one dream, for example, I was working at a telephone job (as I do now) with no hope of doing anything else. (No vacations, no breaks). Or events happen which are constantly disappointing me, etc. I keep thinking about this parallel universe as a cursed place much like 20th century earth! Things look good, very expensive and I feel I am trapped in a never-never land run by Ronald Reagan with the greed of the 1980's where I am never free of material bondage and forced to accept everything "the way it is". (And buy everything to avoid looking poor.) I have actually awakened crying! But I am told to look closer and when I do, I see that everything is rotting away, from the bottom up.

The foundations of these beautiful houses are filled with crumbling concrete, rotting wood, termites, and herds of vermin! There is a darkness about this universe, but it isn't threatening. I feel somehow protected. It is more disappointing.

Sometimes Steve's time travel incidents serve to reassure him that if he does what is righteous or good, that he will be protected from the ravages of this world. A case in point concerns his penchant for gambling. On a recent trip to a place where gambling is legal, Steve crossed over a vortex inadvertently. When he turned on the car radio, he was hearing a program he had heard two weeks earlier. That occasionally happens in any programming situation, but as he kept going, he soon discovered that **all** the programs were ones that had been broadcast two weeks earlier! Thinking that perhaps he had gone back two weeks in time, he remembered that a man had won $1,000 at a particular slot machine at one of the casinos exactly two weeks earlier. He went back to that particular casino, hoping for the same results! But to show the Lord that he had more trust in Him than in gambling odds, Steve invested only a small amount in the slot machine and he won $500. This was exactly half of what the previous lucky gambler had landed! The Lord, he explained had not wanted him to be too successful at what was basically a bad habit! My comment was that the Lord has a great sense of humor!

One of Steve's favorite sayings is "never believe what you hear and only about half of what you see." Steve has warned me many times that "we mistake as reality something that may not been real.

For example it could be an alien hologram!" I asked him to give me a concrete example and he revealed the name of a restaurant with a mirror hanging behind the counter. "I can look into that mirror and see people behind in the back room that are walking in and out of the place where there are no doors!" he said. This made me think that anything tangible and even money is illusory. I recalled how Whitley Streiber had seen his bank account drop to nothing and go back again, apparently the doings of some of his alien contacts.

Steve told me once about how someone had proven to him that they had traveled backward in time by bringing a newspaper back. But after only a few days, it aged, and within a week, it had begun to disintegrate! "If something goes beyond its lifetime, it disintegrates, but if you take an old object like an antique and take it into the past and then return with it, it will be rejuvenated. It is the same with humans, when they time travel they are regenerated. But there is no automatic return loop for a jump through time. Time travel is based on pre-destination and those who are supposed to time travel have automatic time loops set up for them," he explained. But who knows if they are supposed to time travel? "In 2012 AD civilization is wiped out. Or it could be just a barrier, but if Christ doesn't intervene, it means destruction. It could even be the beginning of the Tribulation. If people are able to recognize the Anti-Christ (as Kathleen Keating was able to do when he visited her in her backyard) they might want to try and time travel out of here!" Steve said.

In the May 1997 issue of Magical Blend Magazine there was an article by John Chambers, entitled "THREE WHO SAW THE FUTURE".

In the section of the article about Whitley Streiber he quoted that author as saying "time travel is not beyond the realm of possibility any more.

Even Stephen Hawking has now admitted that time travel is possible and it could be within 10 or 12 years that people will be capable of traveling in time. (Stephen Hawking is one of the most conservative among the good physicists of this world.) I think it is possible to do it now with the human mind which is a lot of what THE SECRET SCHOOL is about." Steve Gibbs would certainly agree.

HOW THINGS APPEAR IN ALTERNATE UNIVERSES (Some additional information)

Steve received a book about 20 years ago. It was given to him by his "other self" and was dropped off at one of his favorite places of recreation. Steve has long speculated that the book came from an alternate universe because it has no publication date, no copyright date or Library of Congress number. This was a book that introduced him to time travel and not only opened up the world for him, but the universe as well. In fact several **trillion** universes (but that's a bit mind boggling for the novice to grasp). In his reports (included in the back of this book) you will read about the techniques he uses to engage his **Hyper Dimensional Resonator** (and other similar devices) to access these other universes. Here are some interesting things about them.

1. There are heaven and hell realms in **all** universes, but the higher you ascend and more enlightened you become, the less likely you will require punishment. There are levels below this one, but they require constant struggle just to survive! (Sounds like we might be getting closer to that perhaps).

2. Steve has discovered that there are universes inhabited by dogs, cats, and by mysterious and dark entities. Some of these include trolls and goblins, even odd appearing beings with stranger still animals and houses. They have histories eerily similar to ours, but with other outcomes. For example there actually is a universe where Nazi Germany won World War II!

3. There are universes, which parallel some of our most cherished fairy tales. For example Steve wound up in a universe where everything appeared normal, but was made of cakes, candies, and gingerbread! (It sounded too much like Hansel and Gretel to me, I would have run and Steve did!)

4. Sometimes, 2 universes may be so similar that only 1 thing sets out the fact that such a switch has been made. For example, one of my favorite old movies has always been **Shane** starring Allen Ladd. I have seen this movie so many times that I can tell you who says what and when they say it! But one evening last spring when I was watching it on a movie classics program, the characters reversed in several scenes. Van Hefflin who played Joe Starret in the movie gave a speech that Shane had always given and there were other

"changes" too. Despite that fact most of the movie remained the same and we saw Shane riding off into the mountains at the end. Before all of this started, I had heard the **pop** and **thud** that switches one over into another universe albeit for a short time!

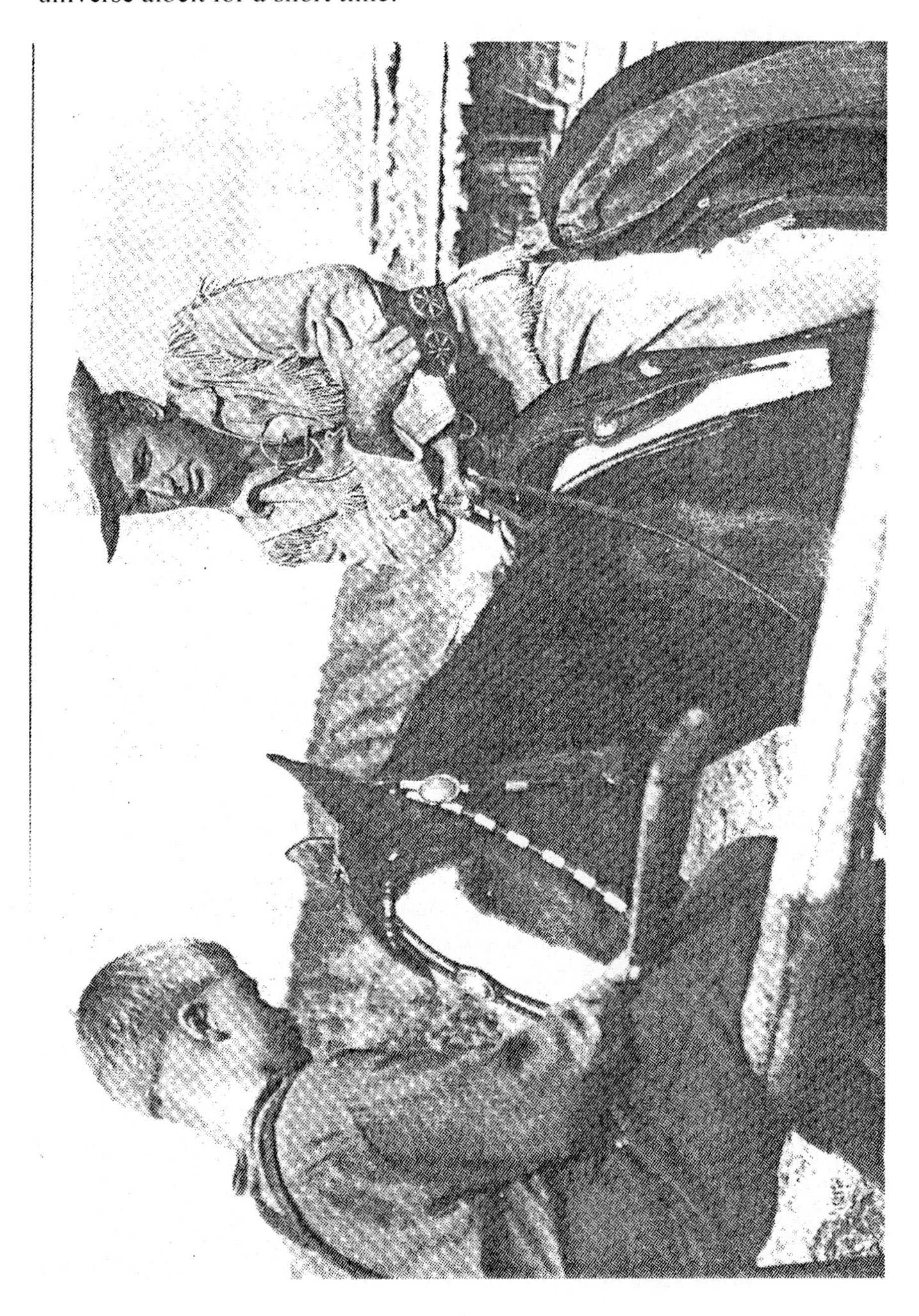

One last comment I will make and this refers to the UFO book which I have written, researched and published over the past 4 years. **Many** of the people I have interviewed about UFO experiences have had experiences similar, if not identical, to Steve's.

This is especially true with the alien's ability to control the flow of time, faster, slower, or even freeze! Steve has, indeed, run into some of the more familiar entities most of us have heard about.

Just for the record, Steve says there are *at least* **20,736,000,000,000** universes available for us to access, and time travel is not new! One of the oldest time travel devices was the **Ark of the Covenant**, the subject of an Indiana Jones movie. "There are instructions in the Bible about how the ancient Hebrew tribes used the Ark," Steve explained. These instructions made him believe that the Ark acted as a *Tesla Coil* and opened a doorway to the Lord. But, as with all time travel, if the wrong person opened the doorway or accessed it improperly, they were history. Some believe that the Ark is hidden and guarded in Ethiopia today. Finding it could prove to be a telling experience, but the Bible states flatly that it shall not be spoken of again.

TO HELL AND BACK - WITHOUT AUDIE MURPHY

The concept of hell has haunted many of us down through the centuries of time. I have heard everything from the traditional horns and pitchforks to the comment that hell is a garbage dump in Jerusalem! But Steve insists that there **are** places of punishment in every dimension!

"Hell seems to be what you fear most," Steve explained to me. "I was taken deep down below the earth and heard the groans and crying of people suffering in great pain. I smelled rotting flesh and an angry hateful voice told me that people were going there by the truckloads, even on a daily basis! There were doors on all sides of you and you had to be careful which ones you opened! Since my weakness is gambling, I was shown a casino, which never closed. Everyone there was sitting in front of a slot machine and winning. When the numbers came up, they put their hands down to grab the money coming out of the machine, but instead had their hands stung by hundreds of scorpions! I could not believe the pain I saw on their faces and the agony I heard in their voices!" he said.

One thing that troubled Steve a great deal was the notion that people could go to hell out of ignorance. "People who say they didn't know it was wrong to do this or that can still wind up in hell," he noted. Apparently even in the other worlds, ignorance of the law is no excuse!

A few years ago someone sent Art Bell a tape which was recorded at a great depth beneath the earth. Apparently some geologists were digging for something and had a tape recorder on. Bell played this over the air and told his audience that it was theorized that this was a tape of hell. It certainly sounded like it. Another AM Coast-to-Coast Show guest, Kathleen Keating, said that hell was deep within the earth. In fact she joked that if Pat Boone **really** went to the center of the earth, his white bucks would get pretty dirty! (An obvious reference to his part in a move of that same name.)

Then, if hell is our greatest fear, what is heaven? It is our greatest wish multiplied thousand of times over! It is a warm spring day, beautiful flower with petals which melt on your lips into the sweetest butter you have ever tasted. It is every moment of happiness you have ever known and every joy you have ever experienced! It is personal, no harps and clouds or pearly gates, unless of course, such images would fit into your concept of happiness!

So where do aliens, angels and devils fit into all of this? Steve can tell that aliens inhabit the bodies of certain individuals. He told me, "they have dark auras, Pat. You can't miss it." To give me an idea of what he was talking about, he sent me a copy of John Carpenter's They Live a short movie that Steve claims is about 98% true. Like many UFOlogists, Steve believes that many of the "ungodly" aliens are reptilians from other dimensions in other star systems. He agrees with David Ickes who has written that the ruling elite families of this world are of reptilian blood and earn and keep their power through the practice of Satanism. During interviews with the various contenders for the American Presidency, Steve noticed both men and their wives with reptilian eyes and forked tongues (does it surprise you politicians have forked tongues?)

Kathleen Keating also asserted that our government knows the way to hell. She asked once, "does it surprise you that our politicians know the way to hell?" She said it half-kiddingly, yet also half-seriously.

What is interesting about Steve is that he has reached many of the same conclusions about aliens and other inter-dimensional entities, as have great scientists like French astronomer, Dr. Jacques Vallee, Dr. John Mack, and some of our astronauts. Yet Steve was not at all familiar with any of their writings!

SOME FREQUENTLY ASKED QUESTIONS

1. **How did you get interested in time travel?**

   A. I was contacted by one of my other doubles or counterparts from an alternate future universe. This was back in 1981. I was given a letter written by a person who was contacted by this other self. This motivated me to start doing research on time travel.

2. **Are there different kids of time travel?**
   A. Yes, there is physical time travel and that is where you can move physically backwards and forwards through time. Then there is quantum time travel. That is when your soul can move into one of your other counterparts in parallel universe.

3. **What is the power that your device generates?**
   A. Pure white light energy-tachyon energy!

4. **What is a tachyon-at least in terms of how you understand it?**
   A. It is a time particle that has multi-dimensional properties, which allow it to move through time. There are positive tachyons that allow you to go into the future, negative tachyons that allow you to go into the past, and neutral tachyons that allow you to lock onto other dimensions.

5. **How does it work?**
   A. It is based on Radionics. It takes your soul energies and steps them up through the zero vector, then converts them into two points of resonance. Whenever you have these two points of resonance, you get a time warp. The electromagnet acts to transmit the soul energies. You then place the open end of the magnet over the stomach region (chakrah) which conditions your soul and aura for time travel. This programs your soul to initiate commands.

6. **Did you say this device had other uses besides time travel?**
   A. It can be converted into a Radionics machine to be used for healing, increasing your psychic abilities, and tuning into the Creator.

7. **Did you say that to travel physically through time that this device needs to be activated over a grid?**
   A. Yes, you've got to be near a grid point. That is an area where the gravitational way lines intersect each other...but you need to find the ones where the U.F.O.'s have been to have it all work.

8. **Without going into all the dial-turning, etc., what is it like once you actually begin to travel through time?**
   A. Initially you feel a great earthquake. Then a huge white blinding light will surround your body. Then things start fading out around you and you start drifting through a void. After you drift awhile, things start forming around you and you find yourself at the time you programmed into the unit!

9. **Can the future be changed?**
   A. There are things that cannot be changed; they are pre-destined to happen! When you **CAN** change an event, you shift yourself over to another universe that corresponds with the event you altered!

CHANNELING DEVICES

Recently, in the fall of 2000 ad. Steve sat down at his word processor and began writing about and designing a new time travel machine that he calls the **Bloch Wall Device**. This is included in the back of this book. He told me at the time, “Pat, this just comes to me, it scares me!” (It scares me too!)

Please note some of Steve’s other reports at the end of this book. These are all recent additions to his time travel reports and contain letters, incident reports, and other tidbits of information that should pull everything in this book together!

DO EVEN THE DEAD VISIT OMAHA????

One might ponder the question of whether the apparitions seen in the OLD MARKET section of Omaha were actually time-slippage or two different dimensions interfacing with each other OR the same things of which ghosts are made! But ghosts of the dead don’t sit down at tables and eat meals, or do they?

Another of my many artist friends told me a story that doesn’t really fit into any neat category, or possibly fits into all categories - you decide!

“Allison” was putting up a display in one of the many quaint little cafes in the OLD MARKET area of Omaha. Since she was lugging around a lot of oil canvasses, etc., she was dressed quite informally. Nevertheless, she was wearing an unusual shade of purple upon which several people had commented favorably! While she was struggling with the hanging of a large unruly picture, she heard a voice from behind her say, “you sent for me, and here I am. How can I help you?”

Totally confused, she turned around and noticed an elderly native-American man standing near her. Although his hair was in long white braids, his face was youthful, even ageless! But what caught her attention was the fact that he was wearing an outfit in the same unusual shade of purple as hers! Later on other acquaintances that had been in the café recalled having seen her talking to the man, so in that sense he was very real. When she agreed to go with him to another place where they could speak privately, he seemed to know his way around Omaha, especially in those parts of greatest interest to native-Americans.

Allison and her Indian friend (who introduced himself as Dr. with an Indian last name) talked for the entire night until she had totally lost track of time! He told her from which tribe he had come and she remembered that at the time she wondered what someone from that eastern tribe would be doing in Omaha. They discussed many spiritual and metaphysical matters that had been concerning Allison and she never did understand how she could have ***sent*** for him or asked him to come and visit her. When they said good-bye at dawn, she had the impression that he was going to Macy, Nebraska, but she didn't know why she believed that, especially when he had not specifically mentioned where he was headed. Later on, while talking about this strange experience with a female friend of hers who was from the same tribe/nation as her visitor, the friend seemed flabbergasted and asked Allison, "do you know who Dr. - is?" Allison shook her head and her friend said, "he is a highly-respected physician and tribal elder…but he has been dead for over 10 years!" She then showed Allison a picture of the tribal elder, and he was, indeed, the man with whom Allison had spent the entire evening! Do the dead, indeed, travel on the wind? If so, from where do they come? Steve's comment on this was that the wise tribal elder had been sent from another dimension or level of spiritual reality to pass on knowledge useful to both Allison and her friends. So people can and do return, but what about events?

When I lived in the South, I often heard stories of people who had seen or heard battles being replayed over and over again.

What was so unusual about that? I told myself, Southerner's were **still** fighting the Civil War, but how about a singular traumatic event that occurs over and over again? Let me tell you about SPRING RANCH, another case of time slippage in Nebraska.

In south central Nebraska not far from the smaller cities of Hastings and Grand Island is a tiny town of Sutton. This is the location of a countywide newspaper system for which I worked when I first moved to Nebraska. If you head south of Sutton on any of the main highways leading to Kansas, you'll run into more small towns like Clay Center, the county seat of Clay County. If you go up to the 2nd floor of the courthouse, you will find the Sheriff's Office and behind the main counter sits an old safe, the kind and color which you have probably only seen in cowboy movies.

If you're polite and ask the secretary behind the counter if you can see ***the noose***, she'll unlock the old safe and pull out an old wooden box. Inside you'll see a noose, or what's left of it after nearly 100 years! Then you might want to ask why is it there and she'll smile and shake her head. She won't know ***why*** it's there, but odds are she'll know the story behind it!

In the late 19th & early 20th century there was once a ranch in those parts of Nebraska. It was called **Spring Ranch** and wasn't too far from where the

Oregon Trail had passed through Nebraska. Spring Ranch was owned and run by a ruthless unmarried woman who was named, ironically, Elizabeth Taylor!

Many stories have surfaced about her over the years that may or may not be true. But the story involving the noose really happened and here is the shortest version possible.

Apparently Ms. Taylor knew more than she let on because an awfully lot of her hired hands seemed to drop off the face of the earth! Pinkerton detectives, law enforcement officials, heart broken sweethearts and lawyers as well as anxious parents formed a regular brigade leading to Spring Ranch in search of loved ones who had given the place as their last known address. Somehow, someone had figured out that Ms. Taylor, to avoid paying their salaries, had poisoned her employees with Paris green (a form of Arsenic, which could be disguised by placing it in a sugar bowl). Ms. Taylor then had the "remains" done away by using quick lime pits when well hidden buried spaces were used up!

But the farmers and ranchers of that area were greatly frustrated because they could not **prove** what they knew was going on! That's when they took the law into their own hands! An old retired rancher from the area told me "*the rest of the story*" and here it is:

"I remember the night they hung her just like it was yesterday! My uncle came for my daddy and I watched them ride off with the white flour sacks over their heads and the two round eyeholes cut out. I was afraid the Sheriff or Marshall might show up so I hid under the bed.

Later when it was all over, I heard my daddy telling my momma what they'd done, they had hung 'er from that bridge, the one outside of town that crosses over the river." He explained the location and then continued the story.

"We kids were all told that if anyone came by and asked any questions, that we didn't know anything! In fact, all of my friends were told the same thing! It seems the populations of two or three towns and villages had gone to the bridge that night for the hangin', but you know what? They say that on certain nights you can still see her handing from the bridge, dancin' in the wind. They say, you can still hear her groaning, and that's a fact!"

Well, he told a good story, and I **did** have to drive over that bridge to get home that night. As I drove along I watched the full moon slip behind the clouds. I floored it all the way home!

So perhaps, we see a possible hell-a-time loop repeating over and over again! Never-ending, tormenting, like Sisyphus rolling the rock up the hill and never quite reaching the top! Maybe that is how Elizabeth Taylor feels at night when she is forced to once again, dance on the wind!

So when Elizabeth Taylor returns to dangle and dance from the end of a rope, or the tribal elder returns to offer sage advise to someone in need of advise, the question still remains; how do they get from there to here? "Allison" had not even heard of Steve Gibbs when she received her visit from a trial elder.

Also, not many ranchers had electricity to activate the HDR units when Elizabeth Taylor danced at the end of her rope. So how can one get from here to there without the use of electricity, or without some kind of instrument for help?

Let's go back to my late friend Quay Fitch. On one occasion, a particular Sunday afternoon, my husband and I were introduced to a railroad engineer from Omaha named J.C. J.C. is a very intelligent fellow and at the time we met him, he had been delving into automatic writing. He was advised that on a trip to western Nebraska he would be shown how entities move from one dimension to another. This came through the automatic writing. The next time he was in western Nebraska, things happened in such a way that his corresponding *"friends"* were able to show him what they promised. "I had stopped the train and gotten off to check something. The other fellows had gone to other parts of the train to check on some other things and then, when I was alone, I heard a voice inside my head tell me to look up. When I did, I saw their *"ship"*. It stayed for a few seconds until they were certain I had seen it, I guess. Then it slipped back into the sky like a letter slipping into an envelope. It just slipped in."

J.C.'s description fits something my husband, Fred, has long observed about UFO's. "It seems almost like they are ship-wrecked," he said. "They come and then they go, never staying very long. It's like *"uh-oh"* we're here again, kind of thing..."

Lastly, I'd like to also comment on my continual dreams about things "*rotting away.*" In her book, **Spirit Song**, Mary Summer Rain revealed how her friend "*No Eyes*", the old Indian Shaman, told her that things would be falling apart. She suggested that planes would be falling from the sky, busses would be crashing, etc. Well it is all coming true! It dovetails with what my dreams have been telling me! For some reason, the astral plane is where I am given my most acute and urgent messages. Maybe we should **all** pay more attention to our dreams!

DEDICATION

I would like to dedicate this report to
Jesus Christ who is my best friend and eternal Savior!

(INSERT PICTURE TITLED "THE 10TH KEY TO THE RIDDLE OF TIME, BY STEVEN GIBBS)

HERE

INTRODUCTION

Even though this may not deal with the subject of this report, many people have asked me, "Do you believe in reincarnation?" Well, my answer to this would be, if God allowed reincarnation to exist, then this would violate the word of God. Because in the Book of Hebrews, Chapter 9, Verse 27, it states:

AND IT IS APPOINTED UNTO MEN ONCE TO DIE, BUT AFTER THIS THE JUDGEMENT.

So, if a person thinks he has lived before, then it can only be one of 2 possible categories; either the person is possessed by a devil, or his soul is linking up quantumly to his counter-parts or doubles in past, present or future parallel universes.

It is interesting to note that the *Theory of Reincarnation* originated from some monks who lived in a monastery high in the Himalayan Mountains in Tibet. However, what people don't realize is that this Monastery is built over an ancient city, which was constructed by fallen angels, which rebelled against God at the dawn of creation. There is a lot of proof to back this up. More so than most people realize. So if you ever decide to take a trip to Tibet, make certain you stay clear of the Monastery, for this place is a trap door which leads straight to Hell!

HOW TO CONSTRUCT THE TA-36 (TIME ANTENNA)

Believe it or not, the Lord gave this information to me when I projected into another dimension after using the **Hyper-Dimensional Resonator**. The theory behind its function is that the double terminated quartz crystal (which connects to the copper ring) acts as an antenna for drawing in the time harmonics through the earth's power grid system. After the energies have been collected, the crystal transmits these harmonics through the copper ring in the form of a vortex where it is then directed towards the target area.

So basically, if we were to build such a device, the chains, which connect the quartz crystal to the copper ring should be constructed out of gold, silver, copper, or steel. Also one end of each of the chains should be glued to the crystal and the other ends fastened or soldered to the copper ring.

When finished, ask the Lord to bless it by praying over it, then sprinkle some salt water over it blessing it in the name of the Father, Son and Holy Ghost. After it has been blessed, all you need to do is mount it on a stand. (See page 2 for details.)

To operate this device, while touching the crystal with your fingers, communicate with it as to where you would like to go. As soon as you feel a tingling sensation, this means that the crystal has received your command.

Next, aim the TA-36 at your bed or where you take a nap. If you've done everything correctly you should be projected in less than an hour.

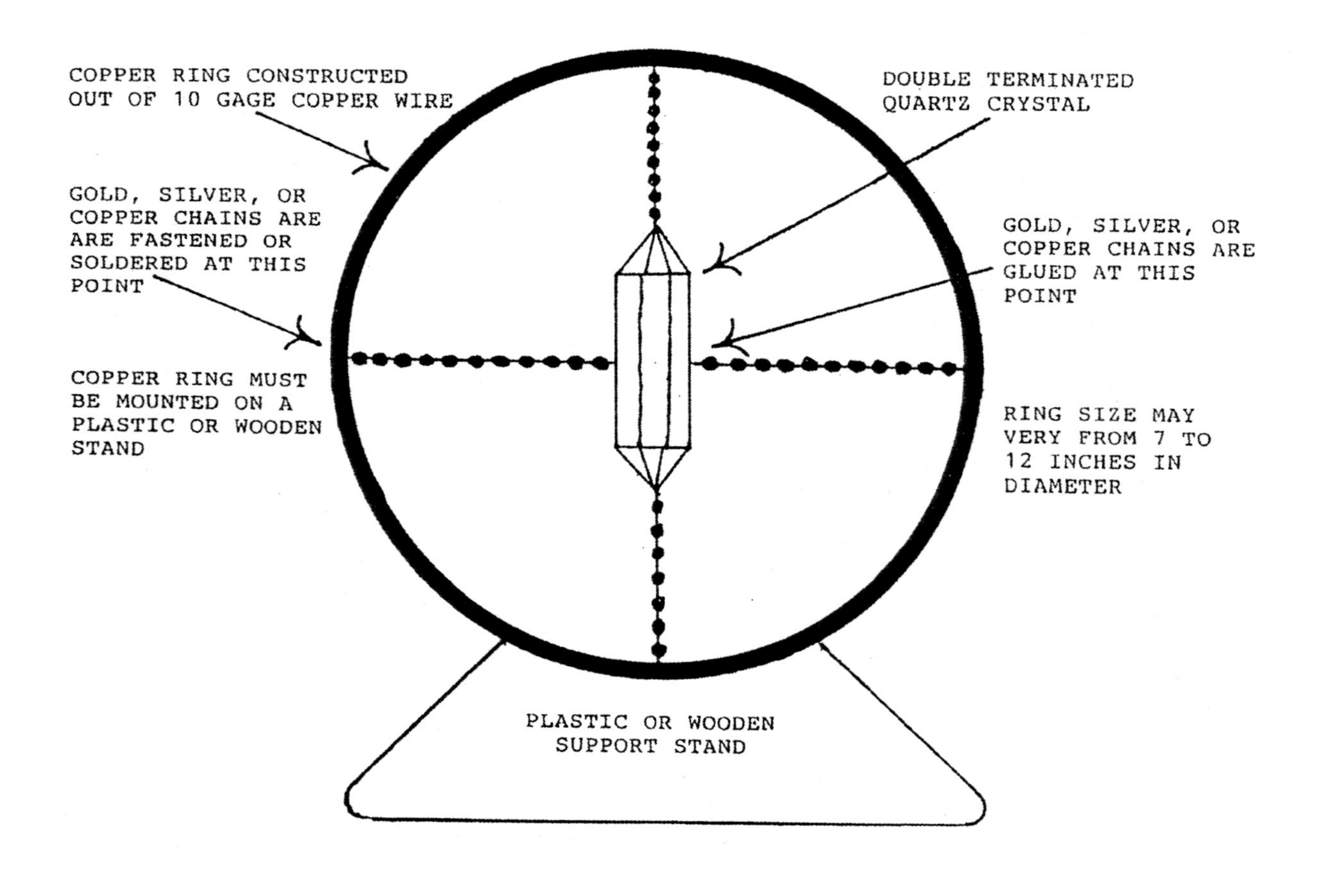
COPPER RING CONSTRUCTED
OUT OF 10 GAGE COPPER WIRE
DOUBLE TERMINATED
QUARTZ CRYSTAL
GOLD, SILVER, OR
COPPER CHAINS ARE
ARE FASTENED OR
SOLDERED AT THIS
POINT
GOLD, SILVER, OR
COPPER CHAINS ARE
GLUED AT THIS
POINT
COPPER RING MUST
BE MOUNTED ON A
PLASTIC OR WOODEN
STAND
RING SIZE MAY
VERY FROM 7 TO
12 INCHES IN
DIAMETER
PLASTIC OR WOODEN
SUPPORT STAND

HOW TO CONSTRUCT THE STM (SPACE TIME MODULATOR)

Don't let the simplicity of this device fool you. If built correctly, it can pack a good punch. As for building the device, pretty much all the parts can be obtained through your local Radio Shack store. The items needed for building this modulator are listed below:

Two 6-amp Rectifiers
Two 1.0K FMD 240V Capacitors
One perf board
One copper clad circuit board
Three 50K potentiometers
Three knobs for mounting on pots
Battery clip for ground connection
One on/off switch
One roll of 22 AWG speaker wire
One project box
One 9-volt battery

Building Instructions:

Both the rectifiers and capacitors along with the 9-volt battery are mounted to the perf board. The 3 potentiometers, along with the on/off switch are inserted and fastened to the top panel of the project box. The 3 knobs are then connected to the stems on the potentiometers or pots. The battery clip is for grounding the unit to either a vortex or grid point. **WARNING**: *Make certain that when you build this device, you use the same purification procedures for the TA-36.*

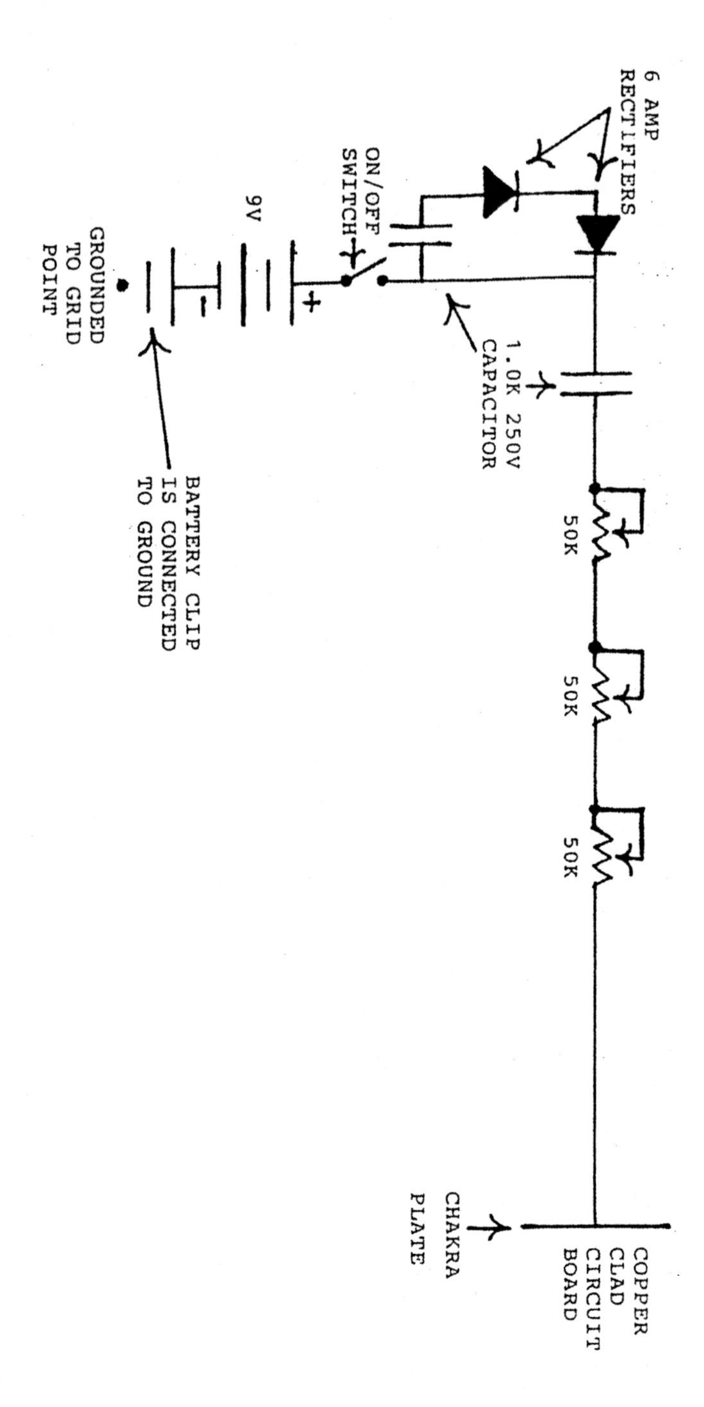
6 AMP RECTIFIERS
ON/OFF SWITCH
9V
+
-
GROUNDED TO GRID POINT
1.0K 250V CAPACITOR
50K
50K
50K
BATTERY CLIP IS CONNECTED TO GROUND
COPPER CLAD CIRCUIT BOARD
CHAKRA PLATE

As you can see on the previous page, the battery clip is connected to the negative terminal on the 9-volt battery and should be grounded to a natural grid point. However, before you do this, make certain that you pray over the grid point asking Jesus to bless it.

Afterwards take some salt water and sprinkle the area 3 times, blessing it in the name of the Father, Son and Holy Ghost. If you fail to do this before you make your connections, a Demon or Evil Angel can come through the vortex and possess your body. *So take heed to this instructions!*

<u>Operation Procedures:</u> to operate this device place the copper clad circuit board (which is connected to the unit) over the stomach chakra or (navel area). Next, while turning each of the 3 dials, concentrate on the following question:

Jesus, what are the rates which will transport my physical body and all of it components to (month), (day), and (year)?

As soon as you feel a tingling sensation stop turning the dial and proceed to the next dial and do the same thing as before. Proceed until all 3 dials have been tuned. Now, all you need to do is activate the "on/off" switch and you're off. If you've done everything correctly, you should be transported in less than 20 minutes. But of course, this will depend on the vortex. **Warning:** *This device can be used for physical, quantum or out of the body time travel depending on whether your actions are pure. All of this depends on faith and the will of our Lord and Savior, Jesus Christ.*

HOW TO CONSTRUCT THE QR-7000 (Quantum Resonator)

This device, like the STM, is extremely simple to build. If you look at the drawing, you will notice that the positive red terminal from the 9-volt battery is connected to a bi-filler winding and one (LED) light. The other negative black terminal is connected to a power switch and a 30-ohm resister. All of these components are then connected to a primary coil, which is wrapped around a double terminated quartz crystal.

NOTE: *The gauge or size of wire in this case is not critical. Anything from 21 gauge on up to 24 gauge can be used.*

As for the secondary wires, these are connected to a microphone cord that is placed around the forehead or 3rd eye region in the form of a sweatband. These microphone cords can be purchased through your local Radio Shack store for less than $10.00. The part number is 278-356.

In reference to the bi-filler winding, this is to be placed directly below the rubbing plate section on your project box. In other words, the box is used as a rubbing plate in order to get your stick reaction. In case you don't know, this stick reaction is usually felt when you stroke the rubbing plate section with your fingers in a clockwise or counter-clockwise rotation. Do this until your fingers stop on the plate.

If you have a hard time mastering this procedure, you may want to try a quartz pendulum instead. This usually consists of a piece of quartz, which is fastened or glued to a copper or steel chain.

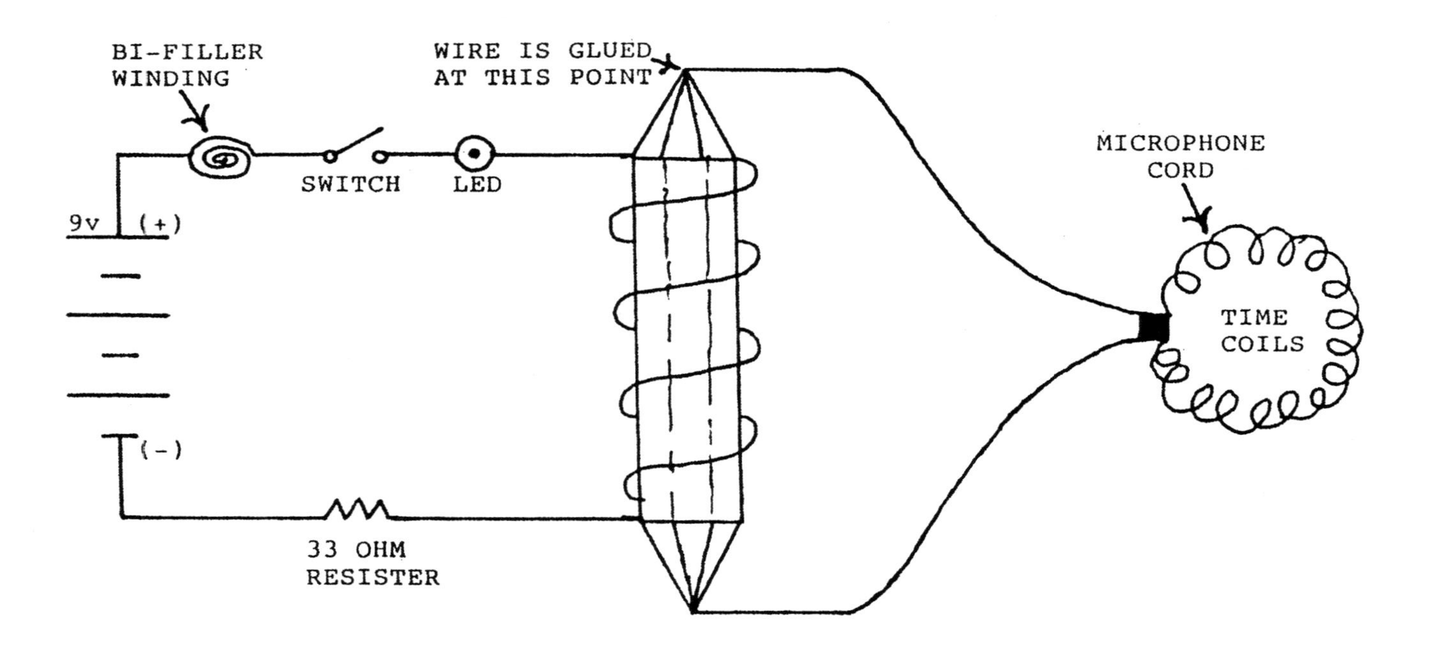

BI-FILLER
WINDING
WIRE IS GLUED
AT THIS POINT
SWITCH
LED
9v
(+)
(-)
33 OHM
RESISTER
MICROPHONE
CORD
TIME
COILS

Use of a quartz pendulum: Position the time coils around your head over the 3rd eye region in the form of a sweatband. Next, while holding the quartz pendulum by the chain over the rubbing plate or bi-filler along with winding, concentrate on the following this command: *"Crystal transport my astral body to (date)."*

As soon as the pendulum begins to rotate, this means that the crystal inside the quantum resonator has accepted the program. Once the crystal has accepted the program, you simply need to activate the power switch and find a comfortable place to relax. If you've done everything correctly your astral body should be transported in less than 60 minutes.

NOTE: *If case you are wondering, the 3rd eye region is located on your forehead directly above your nose. Some people who are extremely psychic usually have a dimple in this area. Also don't forget to ask the Lord to bless your unit. Then with some salt water spray the circuit 3 times in the name of The Father, The Son, and the Holy Ghost. Failure to do this could result in a bad experience.*

HOW TO USE A CRYSTAL BALL FOR TIME TRAVEL

The ancients for seeing into the future have used crystal balls. It is a foregone conclusion that not only can they be used for predicting the future, but they can also be used for time travel as well. The reason for this becomes apparent when one considers the possibility that crystal balls are in tune with all the dimensions.

If you decide to use a crystal ball for time travel, only pure quartz should be used. Lead crystal balls should be avoided as they can send you into a Hell region. Also remember to ask the Lord to bless it then with some salt water bless it in the name of The Father, The Son, and the Holy Ghost. This procedure will purify your ball so that no negative outside forces can influence it.

Next, radionically tune the spark gap setting on a Tesla coil for the year, month and day that you wish to travel to. This is done by holding a quartz crystal pendulum or by using the top of a wooden desk as a rubbing plate in order to get your stick. After the spark gaps on the Tesla coil has been tuned, position it so that the electricity from the copper or aluminum ball arks directly to the quartz crystal ball. Leave it in this position for the space of 20 minutes.

After the 20 minutes, deactivate the Tesla coil and place your hands over the crystal ball. While doing this, gaze into the crystal ball and mentally visualize yourself being transported to that particular time period. If you've done everything correctly you should be transported.

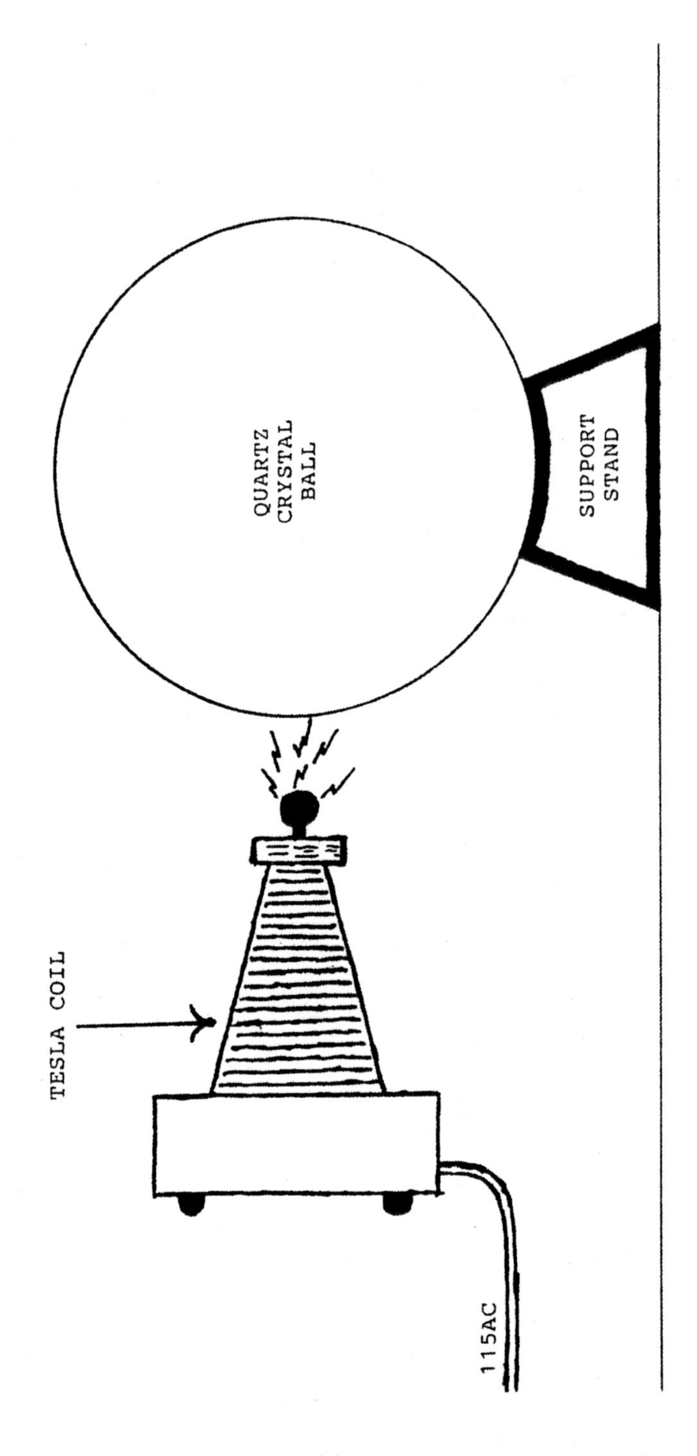
QUARTZ
CRYSTAL
BALL
SUPPORT
STAND
TESLA COIL
115AC

The next section deals with a document account

Of Frank Cuzzo's adventures in time.

His experiments were based on the Multi-Verse Resonator

In the report titled, *"The 9th Key to the Riddle of Time."*

Sad to say, Frank isn't with us anymore. I and my friends feel
That he has either taken off through time or else the Aliens have eliminated him!

Whatever the case might be, I will truly miss him for he was a good friend.

July 7, 1999

Dear Larry:

I want to apologize for taking such a long time to respond to you, and I appreciate your concern. I am very serious about this, and to be quite honest with you I've been looking forward to this opportunity all my life and I'm very excited to be a part of this. I just moved here a couple of months ago and I needed to get settled in. But now I'm working and I have built my circuit again so I am able to resume experiments.

I want to thank you for all of the information you have been sending me, it really helps and is very interesting. Enclosed is a poem I wrote or some say mission statement I wrote some years back that I think you may like, and of course, you will realize that I am deeply serious about this, just as I have noticed about yourself.

Larry, I will try the best I can to give you detailed descriptions of my experiments and experiences. I am not an expert quite yet of course at time travel, but I do consider myself an expert in being able to get around in the other realms which would benefit the ones who know very little. Although I haven't physically time traveled yet, I have quantum traveled and Astral time traveled. Right now Larry, I think until we are able to travel physically consistently we also need to develop our skills in the other worlds, the worlds where are traveling non-physical. I say that because there are many forces, MIB, etc., who will try and get in our way and use those other worlds to defeat us or discourage us, as I have encountered. I will share those experiences with you. Anyway just something to think about if you haven't already.

Here are some of my experiences that I hope you can use:

Sat, 6-13-99 - 6:30 a.m.

I woke up *out of body*. I was able to put hand through wall. I saw a device, which looked like a big lighter about six inches long and two inches wide. It was see through. I pressed the button then heard a high tone, which was kind of bassy. It made me feel a little uncomfortable so I let go of the button. I then decided to press it again, this time I felt myself leave my body, like my soul went out of my head, and I ended up in my apartment again. This time I couldn't put my hand through the wall. I came to the conclusion at that time that I had Quantum jumped into another parallel dimension or something. I heard a knock at the door. It was my Dad, Grandmother, Aunt and Uncle. They came to Colorado to surprise me. Then I popped back into my body.

Sat, 6-25-99 - 8:00 a.m.

Between 5 and 7 a.m. I Quantum time traveled at least I think it was Quantum. No sure what time period or dimensions or universes, but I know I traveled because I was contacted by other travelers. During the course of my travels they tired to contact me several times, but I think they were being tracked, so they had to stay on the move all the time. There were three of them. Two guys and one girl. They seemed a little younger than myself. One of the guys called himself Salsami. He seemed like he had really short hair. The girl had short dark hair, shoulder length. I didn't get her name. They were very excited when they contacted me. I remember seeing them several times but I didn't realize they were trying to get my attention. She tried to give me a way to contact her I think through computer means. An IP address perhaps. I only got half the numbers. 1 557 54...She said she was drugged or on a drug. Maybe that's how they traveled. I suddenly felt an orgasmic feeling standing next to her while she was writing the numbers down on plastic. She said normal paper would just rot away. I was being tracked also. I traveled to another place where I saw myself from above talking on a payphone. My brother was there. He was a little younger at the time. It must have been a parallel universe or something like that cause I don't remember that scene ever taking place in my past. Anyway I tried to talk to myself. I looked him right in the eyes. Maybe he couldn't see me I don't know. I merged with him. At that point I was able to see through his eyes. I felt halfway in and halfway out of his body, but I had full control of his body. I began to cross the street following my brother. I tried to fly but fell on my face. Then I knew I had Quantum time traveled. A car went by…a station wagon 70's type car…like the one on the Brady Bunch…remember? A guy was in it, I've seen him before. Looked like the guy in Ghost, the one in the subway. He was yelling at me that I'd never get back. I heard the name Felix or something like that.

After that I went to the bar, I may have jumped again, not sure but I think I did. I introduced myself to the bartender. A cop came behind the bar and asked for my I.D. I was wearing shorts. I reached into my sock to grab my ID and it was blank. They didn't like that and asked me to come with them so I did. I went into this auditorium where there was a lot of people. I managed some way to get away from them. This was the first time I saw the other travelers sitting in some of the chairs. They tried to get my attention, but I didn't think anything of it. Larry I told you this story from the end to the beginning because that's the way I wrote it in my journal. I hope it makes sense to you. Keep in mind though that I and the other traveler's, were being tracked by someone or something during this time.

Sat, 6-27-99 - 2:50 p.m.
Moonrise 7:32 p.m.
Moonset 4:48 a.m.

Was in a building. Saw men in suits with a woman in the middle. She had short blonde hair and beautiful. They guys in suits tried to attack me but I took care of them. I must have been traveling astrally because I had full use of my psychic powers. The woman was attached by something round in shape, orange in color, it absorbed itself inside the woman and she couldn't get it out. It looked like it was killing her. She had no experience where she was. I told her to will it out of her head. And she did and it left here, then started coming after me. I couldn't destroy it…I had never encountered anything like that before. I asked the girl to stay close to me…because she was in danger…she didn't know how to use her abilities in those realms. A motorcycle cop road up to me. He was a Vampire! Then the whole clan came, hundreds of them. I told the girl to grab onto me and we'll just fly out of here. Well they all had machine guns and shot me down like a bird. I was ok but could feel the hits. They got hold of the girl and were sticking her with a barbed wire type of think in her side. I quickly grabbed hold of it with my bare hands. One of the vampires looked at me in disbelief as I broke the barbed wire with my bare hands, thinking that I didn't have experience to know the power I have in those worlds. I quickly pulled the barbed wire out of her side and told her not to feel the pain. Then after that, the vampires left us along. I don't know what would have happened to her astral body if I wasn't there. I was told I was in the year 1834. I activated the unit at 4:30 a.m.

Larry, this is all I have at the moment, but I will have more for you very, very soon. It is easier for me at this time to type this to you than to send you tapes. I have a roommate and I don't want to talk about these experiences while he is around all of the time. Although I have told him of some...and he will probably be my witness to any physical time travel that I may achieve. I would say though Larry that there may be some dangers in traveling as you can see with the girl in the last story. You can put yourself in danger if you don't know what you're doing. You have to exercise your psychic abilities on those and worlds to let the other beings know that you can defend yourself...if need be.

As for meeting and merging with yourself in the other dimensions or realms or whatever, I don't see any danger in it, unless its physical...which I haven't encountered yet. Still haven't traveled physically yet...but I am working on building an artificial grid point and possibly increasing the power of the HDR Unit. I'm not exactly sure how yet but I'll come up with something. Maybe you can help me with that. Also I need to know if I've built my electromagnet the right way. I used 21-gauge magnet wire for the first 6 rows then used 20-gauge hook up wire, with insulating because I couldn't find any 21-gauge magnet wire. It seems to be working fine but is it putting out the power it's supposed to? Anyway maybe you can help me with that.

Here is a poem some call a mission statement I wrote a couple of years back. I think you will like it.

Guardians of Earth

The choice was man I say
Too bold, no way. This is my home.
My mission, to make sure it is here stay.
I will fight till the end of time
I will fight to keep our planet safe
I give my life to you brothers and sisters
Of planet Earth.
I give my life to the Universe.
I'm not afraid because I know I'm being guided the right way.

I walk with fear as my guide, but not guided by fear itself.
The warrior is within me and exists everywhere around me.
Come my fellow guardians I know you're there, walk with me.
Together, we will take back our home.
Together, we will fight for our freedom as human beings.
Our spirit guides are there for us, and the power of God is within us.
Our future now lies in our hands.
Our hands are the keys, let's use them rightfully
Leaders we are, there is no greater feeling by far
If now, we take control of our lives, our children will become the greatest
Human beings there are.
Let's go now in secrecy. Let's fight the battle silently.
Let's not show off what we do, because of what were doing.
For our works alone, will we be knows!

Written by Frank J. Cozzo
Light Warrior #1

Talk to you soon Larry,

Your Friend,

Frank

December 10, 1997

Steven L. Gibbs
RR 1, Box 79
Clear Water, Nebraska 68726

Seasons Greetings!!

I just wanted to write and let you know I've been using my new Hyper-Dimensional Resonator since the 1st of December, and I am so pleased with how it is working so far.

I chose to work with body heating since I'm an Insulin dependent diabetic. No progress on that particular area so far, but I'm getting some wonderful readings after using the Radionics on body groups. When the instructions said to find out what the vitality was for each group, well...I didn't have anything higher than a 4, and most were between 2 and 3 readings.

After just a few days, I'm getting readings at 65, and 70 and above. Also, in the process I found the correct setting for working with the immune system...a key to my recovery. Since it wasn't listed, I thought you might be interested in my findings, or the setting. It's (60-65). After using it a time or two, my vitality rate jumped from a 4 to virtually off-the-scale at a 10. Boy! Was I pleased! I was prompted to use the setting for 10 minutes, but not to use the yellow switch. I think my body needed a bigger jump-start than most, and I use crystals for healing a lot, which I think adds to the tolerance level. Another thing I've been doing is to chant a Spiritual word while using the equipment. I like to feel I'm protected, and it does make me feel very protected and loved.

Now, I'm getting to the reason I'm writing this letter. What has been so surprising and wonderful gift to me, is that every time I use the unit, I come away feeling a greater awareness of self-esteem. It's like the windows of Heaven have opened and I've come back with a gift of healing that affects my awareness of self and that I'm a valuable person too! I didn't expect that at all, so I'm in wonder about it.

Thank you so very much for being here and sharing your knowledge and considerable skills. It's making things possible for me that I didn't know were there, and I truly consider this a gift of divine love for your fellow man. Keep up the good work!

Regards,

Carolyn Peterson

p.s. I'm sending you a check to cover our telephone conversation in November, I think it was. At least this should help out a bit.

January 19, 1999

Steven L. Gibbs
RR #1, Box 79
Clearwater, NE

Dear Steve;

Enclosed $ for 2 magnets.

Since you are probably the World's foremost authority on time travel, I would hope that some of my experiences may be of some use.

First, I have used the resonator to replace a bridge tooth on my upper gum. The bridge is gone and the new tooth is and has been in place for some years now. Yes, I did have a universe jump during the work on my teeth. I still have several that I wish to replace. Conventional dentistry is not much these days, nor for much of the current orthodox wisdom.

In jumping universe's I did occasion meet a number of the arch demonic who would have possessed me had I given into their wishes. It was quite a trip.

This may or may not been much use for your info, but I hope it is of some value.

Take care, all the best,

Jim Light.

This letter came from a person who used

two Cadius Wound Tesla coils

to travel physically through time.

The last I heard, he took off to another

time period never to be seen again.

Tuesday, July 29, 1997

Hello Steve!

I am sending some of the articles we discussed. The drawings of the device were done on an early (circa 1984-86) application for the Mac, and will either need to be redrawn or translated. Those will be forthcoming as soon as this is accomplished.

I haven't been traveling for a few years as the physical side effects have seemed to become cumulative. I seem to have recovered sufficiently to resume traveling. These effects may have been due to an instability in the field of the device, and phase-translation problems. Al Bielek once said that it might have been a loss of a time lock with the local continuum. I disagree. It's more likely it may be due to a quantum ringing effect from the point traveled to. The farther you go, the greater the amplitude of the ringing. I think the practical limit would be 10-20 billion years in one jump. Then the instability would become irreversible. Smaller jumps with way stations in between for acclimation would be possible. This would be analogous to staged decompression in diving. I wonder if this also applies to parallel traveling, or "sliding". In this case, the residual quantum resonance would cause an unexpected slide after it built up past a critical amplitude.

I keep wondering if those that have disappeared have done so deliberately. If you found a better world, why would you want to go back? If I found a world much friendlier than this one in which my research would not be threatened, would I continue to live here? Probably not. Maybe it's time again to do a little temporal reconnaissance. I think I would come back to offer others a look at how things could be, and then let them decide.

As far as where I have been, there are several time periods:

250,000 A.D.: I first arrived in a desert with a few scraps of brush. It was mid-morning, and the temperature felt like the mid-eighties. The sky was more whitish than blue, which is typical of desert areas, and I did not see any evidence of animal or insect life to begin with. After some exploration, I discovered some underground structures, which were buried by the sand. These were obviously artificial and some of the entrances were holes, which were worn in the walls due to erosion. The people who lived deep within these structures came out at dusk. There were different warring factions or tribes which used what appeared to be directed energy weapons out of crates stored in strategic locations. These crates were so old that the wood crumbled when touched. I estimate that they could not have been older than 5-6,000 years. I could not make out the characters on the sides, but some appeared to be Cyrillic. I looked at one of the weapons, and they appeared to look like a flashlight with a right-angle head. The humans did not seem to have an organized social order, and if left to their own devices, would probably have slaughtered each other in a century or two. I did not study the language as I was spending a great deal of my time avoiding the two hostile factions. I had the impression that they were "transplanted", due to the lack of genetic drift of those which I observed. They were most likely a failed colony which were put there from some other point in time. Maybe Preston Nichols has some ideas on this.

Evidently whatever powered the weapons either had an extremely long lifetime or tapped the zero-point. I tend to think the latter as the exit point of the vortex could have been 'steered" by the energy fluctuations of one of their feuds. My departure was during the eruption of another blood feud which probably caused another disruption of the continuum.

99624 A.D.: I appeared near a city, which was close to present-day Dallas. It was an interesting culture which was based on the concept of permanence. Much of the construction was based on stone and noble metals, which would last for centuries. I felt a certain kinship for these people who had such a keen peace of mind and stability of culture. As far as their philosophy was concerned, it seemed to be a melding of Eastern, American Indian, and some Christian tenets. What most impressed me was the fact that they really lived their philosophy rather than merely espousing it. They called the city Nuel, which I suspect is the corruption of "New El-Dorado". One of the teachers which they quoted was someone whom they called "D'ggan". I was determined to find this personage that they had such reverence for, and which was remembered for over 90,000 years. This place had such a feeling of peace and goodwill that it will be impossible to forget that pinnacle of civilization. If these people had high technology, it must have been completely transparent. The architecture was Egyptian-looking. Everything looked clean though I never saw anyone doing any manual labor. Neither had I seen anything like a robot that did any labor. The only thing I could think of was that it was some kind of matter-hologram, which was refreshed from time to time. I was also impressed at the fact that this utopia actually worked, probably due to the philosophical fortitude of the inhabitants. Maybe I shouldn't have left, but I did. Do I regret it? Maybe, maybe not. I'll always have the memory.

99380 A.D.: Same city, different time. I found D'ggan. I even studied for a while. Absolutely fantastic! Now I understand why their culture lasted so long. I felt a peace that I lacked for so long. It's true that their culture went for thousands of years without needing to use time technology. Maybe that was it. They didn't feel the need for it. They had just about everything else, and they learned to project their consciousness through time, and across it as well into parallel time lines. Some of them here knew of my little sojourn in 99,624. They asked me why I traveled. I replied, "To learn.' They understood. The other students told me that as long as I kept them in my heart, I would always have a way back. I didn't think that was possible from my knowledge of temporal mechanics, but I didn't argue. Those people could do things with their minds that I couldn't have imagined in my wildest dreams. Maybe they kept a way open. Some kind of cross-dimensional portal?

20,000 A.D.: I took a little side-trip before going back. In 99,624 there were legends of "The Fallen", humans who had degenerated into an animal state. They died out thousands of years ago. I guessed that this time would be about right, so I went to check it out. They lived out in the desert away from the city. I found a collection of dirt mounds with creatures about-3- feet tall living in them. They were nocturnal, but if the mound was disturbed a great amount of activity was heard inside with a head poking out here and there and darting back in. These creatures were lemur-like with large blood-red eyes and almost nonexistent noses. Five fingers terminating in claw-like nails for digging and tearing prey apart were also seen. I also saw a skeleton of one of their kind outside a mound picked clean. There were no vultures around, so I assumed they were also cannibals. I saw no evidence that they used or had knowledge of tools. From the skeleton I could see from the structure and position of the teeth that they were primarily carnivores. It's no wonder that they went extinct. With very little food around, they probably fed off each other until there weren't any more left. I made sure that I was gone before sundown. I definitely did not want to be invited to dinner and become the main course!

2008 A.D.: I appeared in Chicago. It was night, and the air was brisk. Fall was in the air, along with a scent which I couldn't quite identify. Something wasn't right. There was a feeling of foreboding that I couldn't -shake. I looked for shelter, and possibly a newspaper. Maybe one that might have been left on the sidewalk, or-dropped and blown by the wind. Things looked too clean, especially for that part of town. A black Humvee went bye with four men in dark or maybe black clothes. I ducked into an entryway and waited for them to go by. I thought it was odd for that part of town to be as dark as it was. Almost like it was evacuated. Then I realized it must be under martial law. This was the south side, and probably the first to be secured. I still wanted to know what was going on. Sometimes curiosity is infinitely unhealthy. So I went on. It was a couple of miles down the road, heading north, that I was spotted by a goon squad. I couldn't dive for cover, they just were on me too fast. I had an automatic weapon shoved up my nose and the other goon demanded, "IDENTIFICATION!" He grabbed the wallet out of my pocket and said, "HE'S ONE OF THEM!" and stuffed it back into onto my pocket. The other one said, "LOOK AT HIS HAND!", while another one muttered, "another friggin' terrorist!" The third one flipped open something that looked like a cell-phone and said, "Another for disposal".

The goons put my hands behind my back and put something that felt and sounded like a wire-tie on them. I heard the zipping sound when they lightened it. Disposal eh? That definitely did not sound good. Two of them t me in the back of the Humvee and the other two stayed behind to look for, more "terrorists". It was then that I realized what that funny smell was, and it wasn't a barbecue. Or at least the barbecue didn't involve chickens. It was burning flesh. I needed a plan. I already suspected what their plan was. I needed one of my own.

The vehicle tore down the empty streets at 60 miles an hour. It stopped at Soldier Field where I saw a line of people with the same wire-ties on their hands. They passed a rope through the loop of the wire-ties and a guard tied the end to a post. I knew if I went in that line that would be the end—so I jumped off the Humvee before it came to a stop, before the guards could react. I couldn't run down the street without getting cut down so I went inside. It was unexpected on their part, and for a few moments they were stunned at the action. I found a lighter on the floor and bent down to grab it, failing down on my back in the process. There was a wad of paper on the floor and I managed to light it. In that position, it wasn't easy. I felt the heat and saw the light of the flame reflected on the opposite wall. The odor of singed hair filled my nostrils as I felt the plastic melt and the wire tie pull free. I burned my hand but there were higher priorities to be taken care of, the most important of which was escape.

Somehow, I found my way into the stadium and was there to see a group of hostages burned to death with flame-throwers wielded by those in the black uniforms. I realized then that I should have taken a uniform to give myself a disguise. Fortunately, the military was too busy at the time to pay any attention to me and I was afforded the chance for escape. It was during a passing glance through a window that I caught the date of 2008. I don't recall what the exact month and day was, but considering the state of mind that I was in at the time, getting that date is enough for me.

There were a few other times which I have had the privilege of experiencing, one in approximately 5000 A.D. and another a few centuries from the present date. Some of the negative experiences which I have had prevented me from pursuing further travel, as well as some physical effects which were undesirable. Why would I want to do this again? Perhaps the threat of 2008 and the sneaking suspicion that whatever effect I may have had in the intervening time, it was not nearly enough to prevent the horror that I have seen from taking place. Perhaps it is the experiences of better times and places, and the promise of a future that may still be there. Then again, it could be that curiosity of what's "out there", waiting to be seen. For whatever reason, I know within my being that the far future is a part of my future as well as the past. It's kind of difficult to explain, something which is felt more on a para-conscious level. Maybe because that when you have gone there, part of you still exists at that time if the theory of simultaneity is accurate. Do you have the sensation of deja vu after 'returning from the future? I do, and it usually lasts 48-72 hours, depending on when I've gone. It's probably due to the temporal resonance which I've mentioned earlier. A somewhat accurate analogy would be skipping a stone on the surface of the water, and in this case YOU are the stone and the destination time is equivalent to the water.

Well, enjoy the papers. They're yours to keep. Give me feedback on what you think, and whatever information you think is appropriate in trade. If you hear about any file translation programs that can open up old MacPaint and MacDraw programs on the Power PC let me know. Meanwhile, I'll either see if I can use a can opener or - if I'm really desperate - redraw them from memory.

Good luck in your work and if I can be of any assistance let me know.

Best Regards

Bill Donavan

NOTE: "Wild Bill" Donavan was head of the O.S.S. before it became the C.I.A.

(TIME TRAVEL CATALOG)

STEVEN L. GIBBS
R.R #1, Box 79
CLEARWATER, NEBR
68726 U.S.A.

Dear Friend,

As you will see on the following pages, I sell an instrument, which can be used, for both out of the body time travel and in help healing the sick.

The name of this unit is called the Hyper Dimensional Resonator. The following pages will show you a drawing of what this unit looks like, describe what the unit can do, and my booklets and reports are revealed, plus information on how to place an order.

If you would like to talk to me in person, my phone number is given in the following manner: (402) 893-3809. I hope to hear from you all in the very near future.

Sincerely

STEVEN L. GIBBS

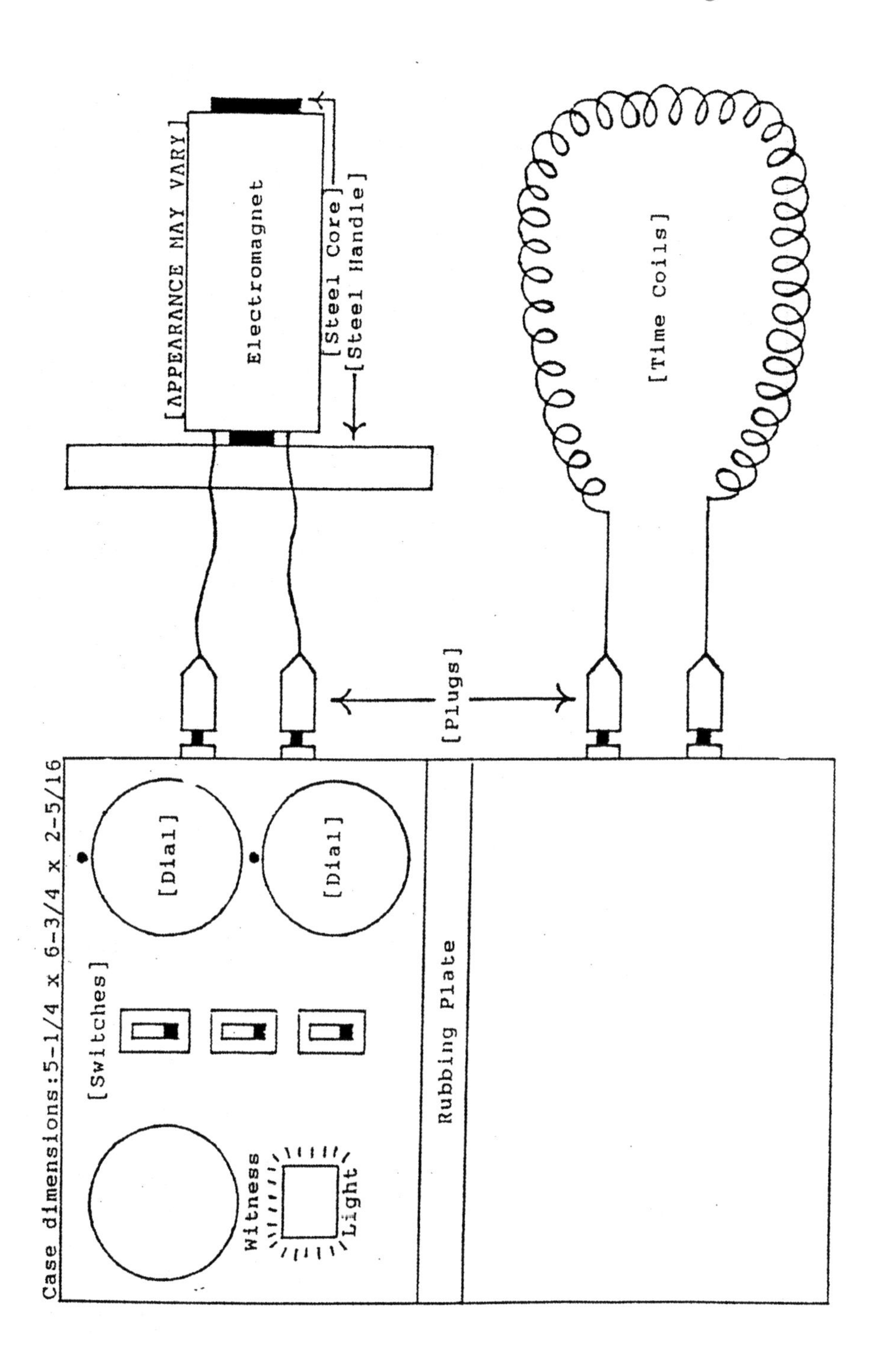
[APPEARANCE MAY VARY]
Electromagnet
[Steel Core]
[Steel Handle]
[Time Coils]
[Plugs]
Case dimensions:5-1/4 x 6-3/4 x 2-5/16
[Dial]
[Dial]
[Switches]
Rubbing Plate
Witness Light

The Hyper-Dimensional Resonator

This here is a two dial, one bank treatment instrument, which plugs into a normal 110v outlet. This device generates an AC/DC, 60-cycle, alternating frequency, which generates an unlimited amount of white light energy. This device comes equipped with a witness well, phenolic rubbing plate, multi-dimensional stabilizer, clear switch, power switch, time coils, and one electromagnet. The nice thing about this unit is, when giving a treatment, you may broadcast the frequency by means of a witness, or you may hook-up the electromagnet directly to the person. Among other things, this unit can also be used for out of the body time travel. As far as the physical aspects are concerned, yes you can use the hyper-dimensional resonator for physical time travel, but only when activated over a natural grid point, or in a place where UFO's are sighted.

Bonus features

Each unit that I sell has a built in electronic sensor coil, which is located directly beneath the rubbing plate. These coils are specifically designed to pick-up and amplify thought induced white light energy. Not only that, but these coils are also designed to help increase the stick reaction when using the rubbing plate.

Operation procedures

This device works just like a Radionics machine. In other words, after the unit has been connected to a 110v outlet, the dials on the machine are then tuned radionically for the year, month, and day that you wish to travel to. This is done while the time coils are positioned around the head. But anyhow, after the rates

The Hyper-Dimensional Resonator Continued

Have been found, the unit is activated, and the electromagnet is placed over the solar plexus for the space of 3 minutes. After the treatment is over with, the unit is de-activated and the time coils removed the person now finds a comfortable place to relax in order for the energies to take effect. If you have done everything correctly, you will then be projected to the year, month, and day. As for the healing procedures, this information is included in my instruction manual.

The hyper dimensional resonator with time coils and one electromagnet, sells for $360.00. Instruction manual is included.

Double terminated quartz crystals

It has been proven that if a double terminated quartz crystal is placed inside the witness well on the hyper dimensional resonator, it will greatly improve its performance for whatever the purpose might be. These crystals average from 1 to 3 inches in length. Price: $12.00 per crystal.

INTRODUCTION

The reason for writing this report is to show people that man does not have to use sophisticated mathematics in order to understand how Time Travel can be achieved. To assume that what is being taught in today's schools is an absolute truth, can only be a lie. For to believe-in such a lie, can only lead to imperfection. But since mankind has chosen this path, the inevitable must surely come to pass.

It is impossible to say where all of this might lead. However I do hope that the people who read this report will use the knowledge wisely. For those who don't, I leave them to their doom. But for those who do, may the Lord bless you

You're probably wondering by now, if I am really an Alien. My answer to this is, there may be more truth to this than meets the eyes. According to the information, which I have received from the Kennedy Space Center, there is a good chance that my parents came from the Star System called Phoenix. If this is case, then this would explain why I have had so much difficulty in adjusting to Earth's own environment.

Perhaps in the years to come, more will be revealed to me on the true nature of my origin. And you know what they say about Aliens? They're weird.

TIME TRAVEL PHYSICS

Before I begin I must first point out, that the only reason why Formulas can be used for analyzing a specific problem, is due to the fact that a Formula helps in focusing the Belief Fields which emanate from the persons soul. So in other words, whenever a person uses a formula or equation, to him this constitutes a reality. By keeping these things in mind, you shouldn't have any problem in understanding what has been written in this report.

To begin with, there are basically 3 sets of equations which can be used for accomplishing Physical or out of the body Time Travel. I have used these formulas quite often when building my Time Travel machines. As a matter of fact, all my research is based on 3 formulas. The formulas which I have used for accomplishing physical or out of body time travel, are given in the following manner:

(THE GRAVITON FORMULAS)

(X, Y = Ø) (X1, Y1 = Ø) (X2, Y2 = Ø)

The Graviton formula, which I have, so diligently labeled, can be used for an infinite number of things. However before we can proceed any further, we must first have a good understanding of what each symbol means. There are infinite number of meanings which can be applied to each symbol, however in order to simplify things, the meanings which I have given to each Letter or Symbol, are listed in such a way so that anybody with a little bit of knowledge can understand them.

(INTERPRETATION FOR EACH OF THE GRAVITON SYMBOLS)

X - Represents an AC Field

Y - Represents a DC Field

X1- Represents a High Frequency Field

Y1- Represents a Low Frequency Field

X2 - Represents a Paramagnetic Field

Y2 - Represents a Diamagnetic Field

Ø - Represents the Zero Vector or Twilight Zone

it may also be referred to as the (PHI Value) which is (1.618).

It is interesting to note, that whenever the (PHI Value) is used in the construction of a material object, a Time Warp usually occurs around that object.

However, in getting back to the problem at hand, whenever the (Graviton Formulas) are used singly or in series with one another, they can be used to help build a device, which can actually be used for Time Travel. This is because whenever the (X) and (Y) Values are combined with one another, we can zero in on the Twilight Zone. And sense the Twilight Zone is in attunement with God, all things become infinitely possible provided that the (X) and (Y) Values are in resonance .

So now you're probably wondering, how can we use these formulas for constructing a Time Travel machine? It's simple. Merely start out by building a capacitor which corresponds to this equation: (X2, Y2 = Ø). This capacitor should be constructed in such a way, so that it resembles the drawing, which is shown below:

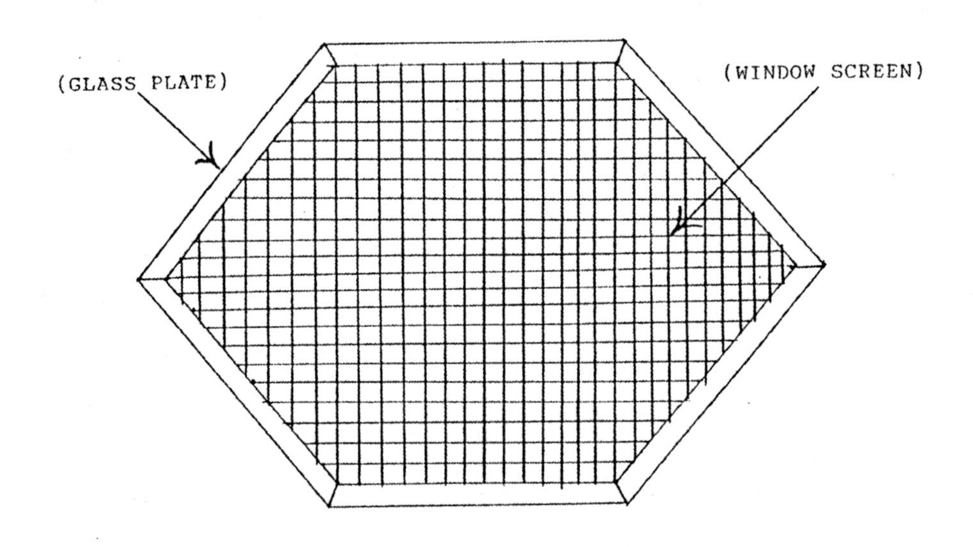

As we can see in the illustration, it doesn't matter too much as to what size it should be, just as long as it is shaped like a hexagon. Also each side of the capacitor should be the same length, otherwise the Belief which is being transmitted from the plate, will be imperfect. This could cause problems with your Chakra points.

At this point it should be noted, that whenever you construct a Capacitor of any type, you are automatically using the formula (X2, Y2, = Ø). And sense the (PHI Value) is used in the formula, the plate itself will develop a Life Force of its own. It should also be pointed out, that sense there is Life Force, which co-exists with the plate, it could now be programmed to do anything you want.

Now in order to program this Capacitor, all we do is stroke the Glass portion of the plate, while concentrating on the following command: (Transport me to the Date and Year). As soon as you get a stick so that you can no longer move your fingers, the Capacitor should be fully programmed. After you have finished programming the Capacitor, place the Screen Side of the plate over the Solar Plexus for the space of 30 minutes. After the treatment is over with, you then find a comfortable place to relax in order for the energies to take effect. If you have done everything correctly, your Aura or Physical body should be transported to the date, which you programmed into the Capacitor. On the average, it usually takes about 70 minutes before the energies will take effect .

***Note**: People who work with these plates are in serious danger of getting possessed by an Evil Spirit or Demon, unless of course, you have a faith in Jesus Christ. Saying a prayer to Jesus Christ prior to any experiment, more than protects you for the forces of evil.*

As you experiment with these capacitors, you will find that some of these plates work better than others. The reason for this probably lies in the fact, that certain sizes or the materials being used, are in resonance with the person's life force or belief fields. Therefore if you run into any problems, it may pay to use a different thickness of glass. When I talk about this, I do not mean to vary the shape of the Capacitor, just the width. It might also pay to use Yellow tinted glass. You may also want to try green, blue, indigo, and violet colored glass. The reason for doing this is simple. Each color corresponds to a different Chakra Point. If is therefore, my opinion that yellow should be your first selection, then if you want, you can always experiment with the higher colors.

***Note**: Never under any circumstances use Red or Orange tined glass. If you decide to use these colors, then you are probably on your own, because not only* *will it lower the frequency rate of your Chakra points, but you could also end up in Astral Hell. And believe me, this is not fun place to visit.*

Once you have found a plate, which can transport your physical or spirit body, there are still even further ways of increasing its effects. One of these ways is by using the second formula, (X1, Y1 = Ø). This formula states that whenever a High Frequency Field is combined with a Low Frequency Field, we can tune ourselves into the Zero Vector or Twilight Zone. To use this formula it will first be necessary to get a hold of a Tesla Coil and one Van De Graff Electrostatic Generator.

Once these items have been obtained, you proceed by charging the plate with High Frequency Electricity and Low Frequency Static Electricity. In other words, after you have programmed the plate using the procedure you must simultaneously zap the plate using the above procedure. You must make sure that while you are doing this, that you do not come into contact with the plate. If you do, you could be in for a Hair raising experience. This voltage hurts! But anyhow after you have zapped the plate for the space of 10 minutes, you then proceed by placing the Capacitor over the Solar Plexus as described. The effects that you get, should be much better than what you experienced before.

The next step to take, is to amplify the Energies even further. To do this, we must-now make use of the 3rd and final formula, which is (X, Y = Ø). This formula states, that whenever an AC Field is combined with a DC Field, a Time Warp occurs which tunes our minds into the Zero Vector or Twilight Zone. Now in order to create these fields, one should try to get a hold of a Hyper-dimensional Resonator. (Read the last pages in this report.) Once you are able to get a hold of this instrument, and after everything has been plugged in correctly, while the Screen side of the plate is positioned over the Solar Plexus, position the open end of the Electromagnet over the glass side of the plate for the space of 3 minutes. After the treatment is over with, disconnect yourself from the plate and instrument, and find yourself a comfortable place to relax. If you use this procedure in conjunction with the first two procedures, the effects that you obtain, will be increased to such a point, that you should be able to do anything you want. You must remember however, that before Physical and Spiritual Time Travel can be achieved, the (X) and (Y) factors in the formulas must be in resonance with one another. This can only be accomplished through the process of tuning your own mind as you would with a Radionics Machine. So in other words, by selecting the correct materials to use, you are in effect creating something, which you physically believe to be a reality. This same rule can be applied when working with different types of Frequency's and Voltages.

Another method which you may want to try, is that after you have programmed the plate by using the procedure given, transmit some High Frequency Electricity from a Tesla Coil into the capacitor while the Screen side of the plate is pointing towards your face. After 3 or 4 minutes, de-activate the Tesla Coil, and while the Screen side of the plate is still facing towards you, take the bare ends of an Extension Cord that has been plugged into a 110 volt AC outlet and short out the bare ends of the wires by touching both of the terminals on each side of the Screen. If you have programmed the plate correctly, your physical body should be transported instantaneously to the date that was programmed into the capacitor.

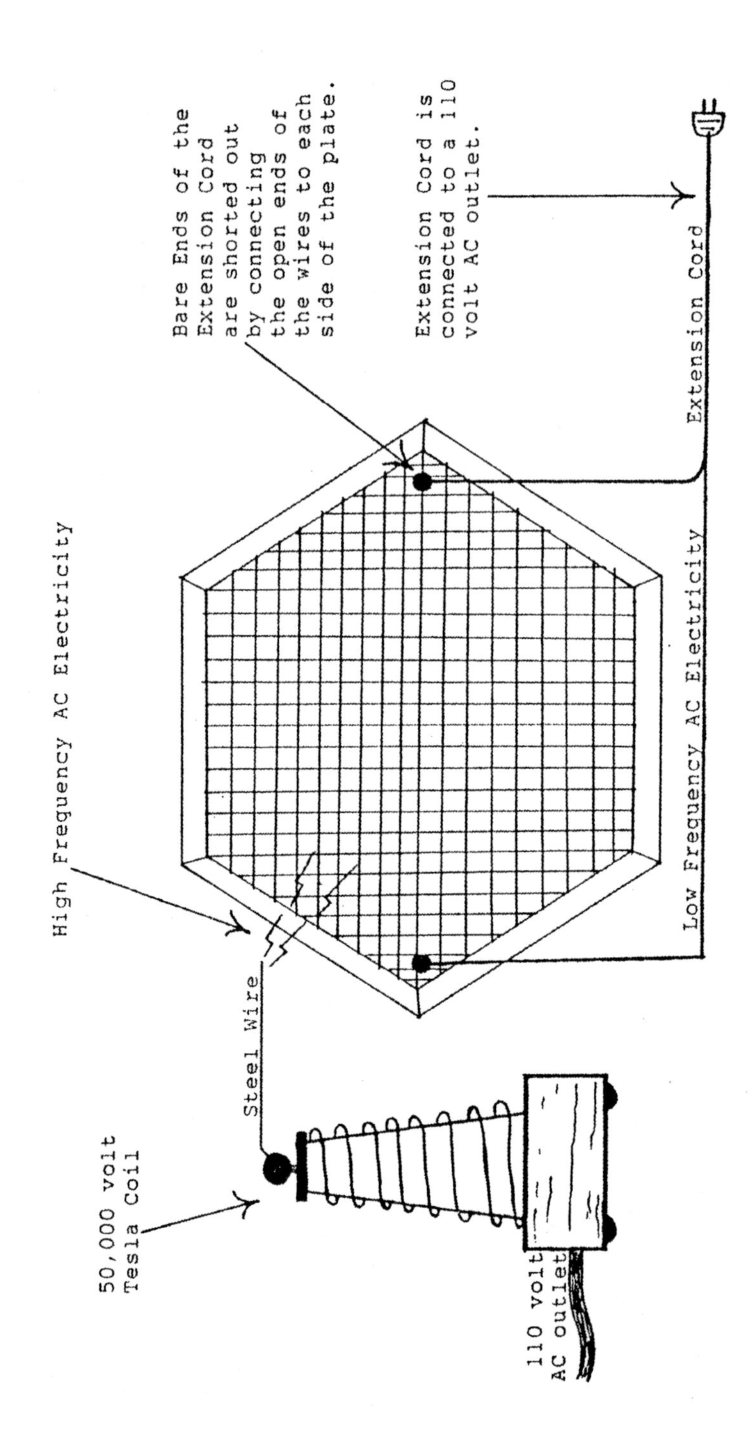
Bare Ends of the
Extension Cord
are shorted out
by connecting
the open ends of
the wires to each
side of the plate.
Extension Cord is
connected to a 110
volt AC outlet.
Extension Cord
High Frequency AC Electricity
Low Frequency AC Electricity
Steel Wire
50,000 volt
Tesla Coil
110 volt
AC outlet

To give you some idea of how all this got started. Back in the year 1981, I was contacted by what I believe to be my other self. The letter which I received at the Sunset Plaza in Norfolk, Nebraska, was dated 1992 A.D. It also had a month on it, which I can't seem to remember. Evidently the letter which I received indicates, that sometime around the year 1992, I will travel back into the past to make contact with myself, just like my other self did in the future. One of the things that the letter mentioned which I can barely remember, is that it said: (The path to the truth can be found in the Pyramid of Giza). It also had a riddle, and it went like this: (The riddle can be solved when 79.613 is dissolved). These two verses holds to key to Time Travel. I have already deciphered the riddle. As a matter of fact it was just last year when the answers came to me. Basically the 79.613 number can be converted into the (1.618) value which is used in my Equations. It is also the same number that was used in the construction of the Pyramid of Giza. If we decipher the riddle even further, we will find that it also reveals that the 7.8 Hz. Frequency, is the Frequency to use for accomplishing Time Travel. It seems that whenever a person travels physically through time, his Alpha or Theta Waves are vibrating at this Frequency. Once that your Brain Waves begin to oscillate at this Frequency, your mind then becomes tuned to the Zero Vector. I have found that this is the only way in which Time Travel can be achieved. So basically what I am trying to get at, the Chronological Time Reflector causes your brain waves to vibrate at 7.8 Hz. Cycles per second. That is all that these units do. However, for some people, they might think differently, but when you get right down to it, it all leads to the same thing. Mind verses matter, and this just barely covers the subject.

Another experience which I had that occurred around the month of Sept. in 1986, dealt with a variation of the Chronological Time Reflector. In other words, after I had finished programming the plate, I proceeded to zap the plate with 50,000 volts of High Frequency Electricity. Sometime after the Experiment, I clipped off one end of an Extension Cord and plugged the other end into a 110-volt, 60 cycle, AC outlet. I then laid the bare end of the wires on top of the screen. Just then I accidentally dropped something on the floor. After I had reached down to pick it up, the bare ends of the Extension Cord which I had laid on top of the Screen, touched the Screen portion of the plate. What was to follow, would be the most terrifying experience I had ever had, because as soon as the bare wires touched the Screen, a massive short occurred. All I can remember is seeing two flashes of white light energy, and the next thing I knew, a white mist had fallen over the entire room. I didn't realize that I had traveled physically through time until I looked at a Calendar the next day. Not only that, but hen I asked lady in Plainview, Nebraska as to what the date was, she said it was the 17th, but my Quartz Watch indicated the 16th. When I had returned home later on that day, I had discovered that all of my Experiments had been dated wrong. Then I finally realized that I had actually traveled one day into the past. To this day, I am not exactly sure as to whether or not I returned back to my own present Time Line. Perhaps I will never know.

Another Time Travel instrument that was sent to me by a man who lives in Osceola, PA., is revealed on the following pages. According to what I have read in his letter, this device can actually transport objects into the future. However, he says that there is somewhat of a Time delay effect shortly after the unit has been shut off. In other words, after the unit has been de-activated, it takes anywhere from 10 to 15 minutes before the object is teleported. I have never built this device as yet, however I plan to eventually.

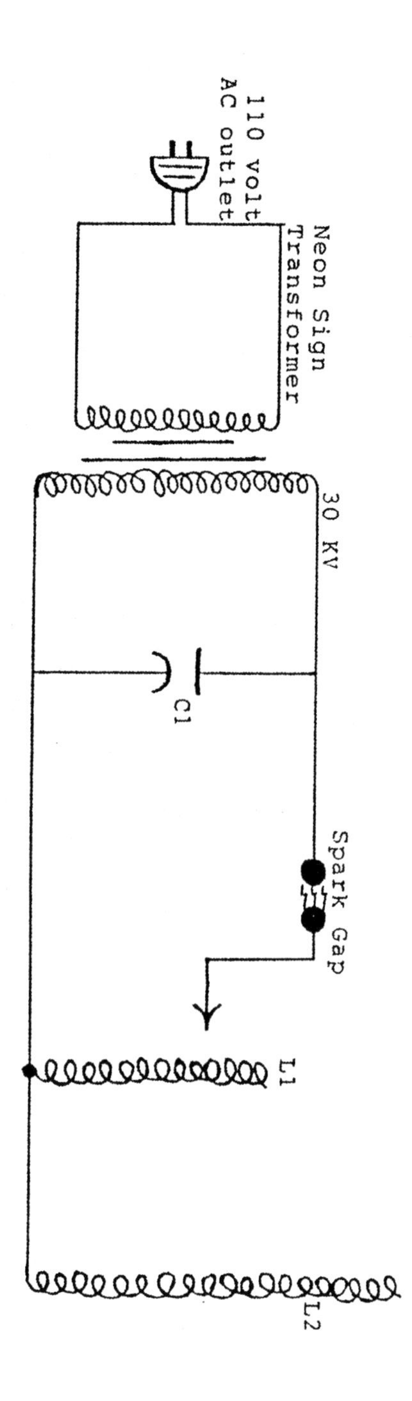
110 volt
AC outlet
Neon Sign
Transformer
30 KV
C1
Spark Gap
L1
L2

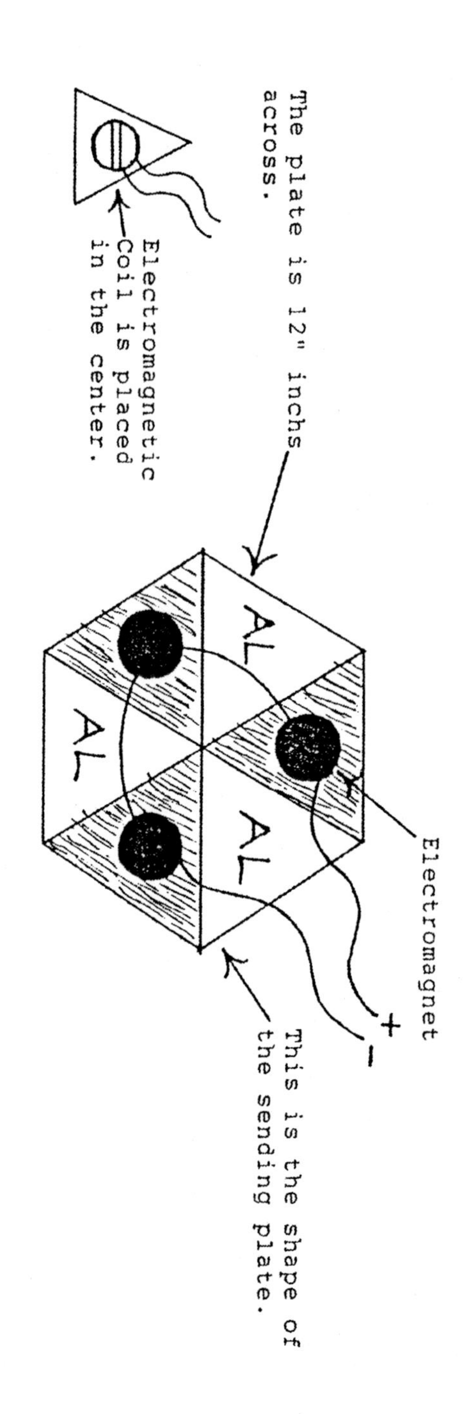
The plate is 12" inchs across.
Electromagnetic Coil is placed in the center.
AL
AL
AL
Electromagnet
+
-
This is the shape of the sending plate.

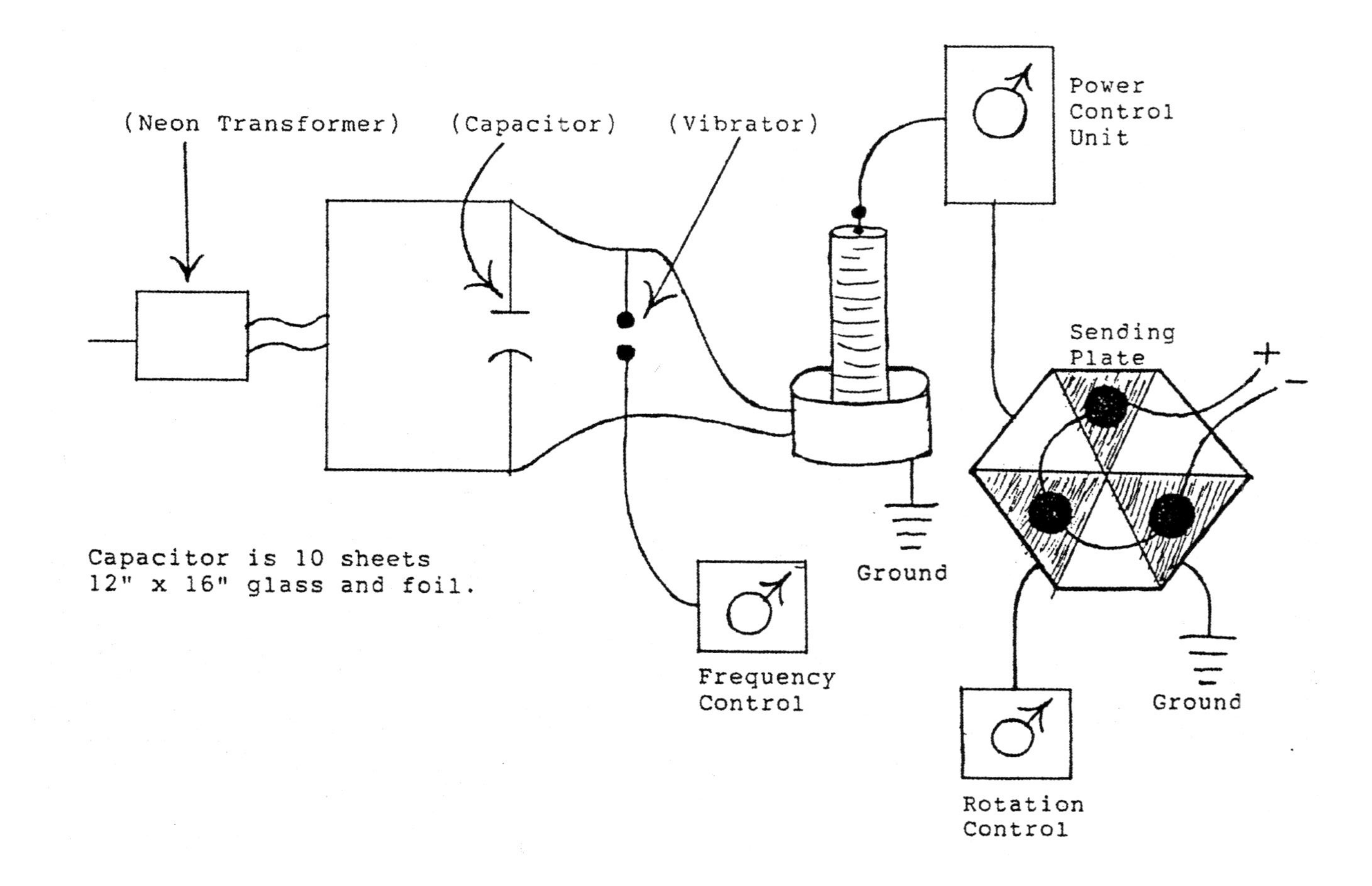
(Neon Transformer)
(Capacitor)
(Vibrator)
Power
Control
Unit
Sending
Plate
+
-
Capacitor is 10 sheets
12" x 16" glass and foil.
Ground
Frequency
Control
Ground
Rotation
Control

Throughout the vast regions of Space-Time itself, there are an infinite number of ways to build a Time Travel Machine. One of these ways is to get a hold of a Rubbing Plate (GSR), or pendulum, and proceed by asking (Yes) or (No) questions as to what type of components one should use in order to build a Time Travel Machine. There are a number of ways of doing this. First of all begin by asking what size the instrument box should be for holding the components. Next, determine what type of energy one should use for achieving these effects. Then you ask what type of components one should use in order to transmit this energy. If you have done everything correctly, you will end up creating a device, which is based entirely on belief. And since you have created a device, which is built entirely on belief, you can be damn certain that it will work! Especially if you use a (GSR) while asking your questions.

For those of you who do not know what a GSR is, this is basically a Galvanic Skin Response Meter. One of these devices may be purchased from me for $150.00. Instructions are included. Please allow 4 to 6 weeks for delivery.

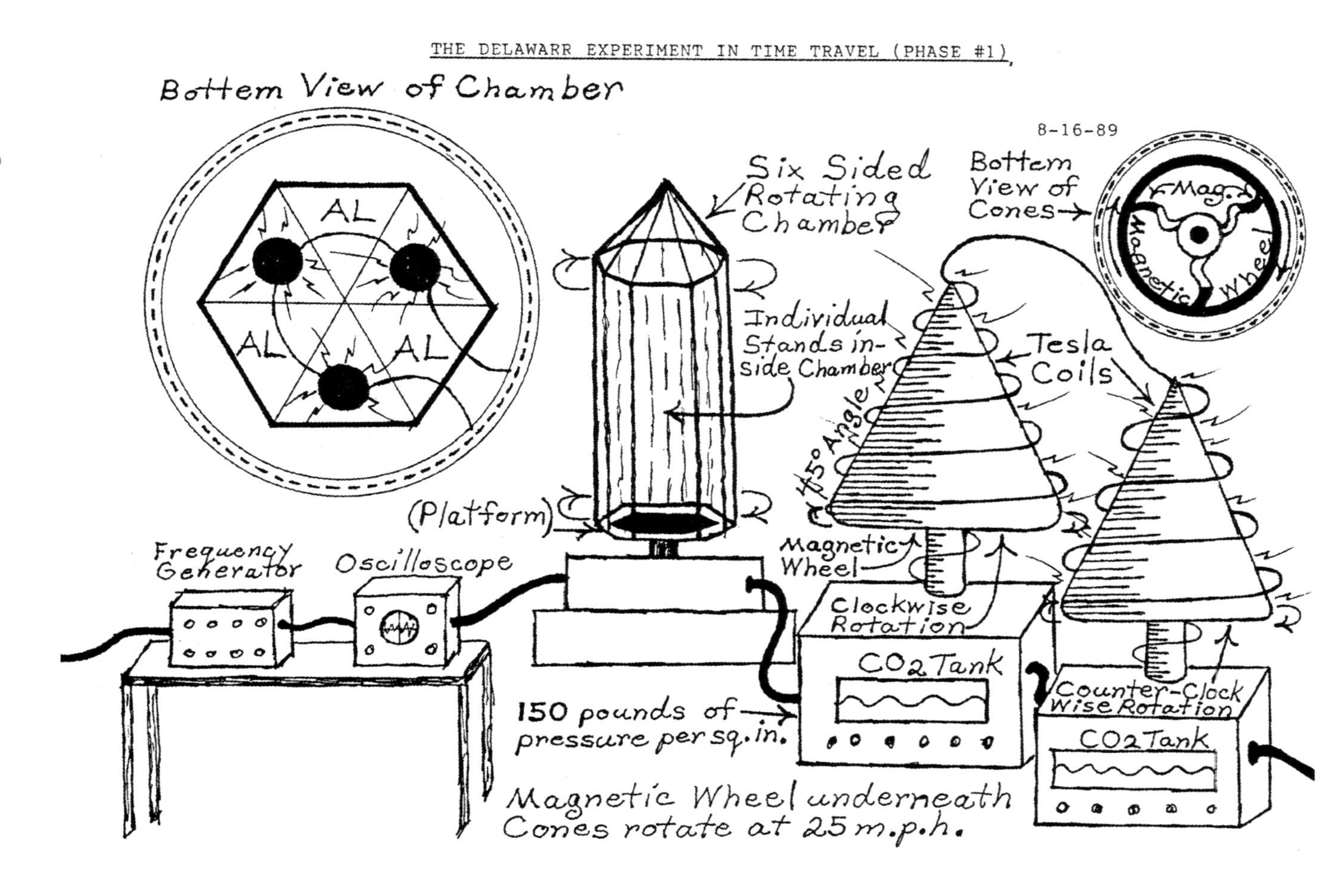
THE DELAWARR EXPERIMENT IN TIME TRAVEL (PHASE #1)
Bottem View of Chamber
8-16-89
AL
AL
AL
Six Sided Rotating Chamber
Bottem View of Cones
Mag.
Magnetic Wheel
Individual Stands in-side Chamber
Tesla Coils
45° Angle
(Platform)
Frequency Generator
Oscilloscope
Magnetic Wheel
Clockwise Rotation
CO2 Tank
Counter-Clock Wise Rotation
CO2 Tank
150 pounds of pressure per sq. in.
Magnetic Wheel underneath Cones rotate at 25 m.p.h.

DEDICATION

I would like to dedicate this report to Jesus Christ who is my best friend and Eternal Savior.

INTRODUCTION

The basic reason for writing this report is to shed some light on the different ways in which an individual or person can travel through time. One of these ways of which most people have little or no knowledge is Quantum Time Travel. This is a subject that most scientists refuse to talk about which is either due to their own stupidity or lack of insight on how God or the Creator operates through this universe.

Now for those of you who are curious, you may have pondered the question, has Steven Gibbs actually traveled through time? Well my answer to this question is, since 1981 when I first started to experiment with time travel, if my calculations are correct, I have traveled to over 10,000 different parallel universes or time dimensions. So one could say in a sense that I am somewhat lost in time.

QUANTUM TIME TRAVEL

There are many different ways to travel through time. Among these different types there is astral, quantum, dimensional, and physical time travel. However for the sake of simplicity we will only deal with the quantum aspect.

Now for those of you who do not know what quantum time travel is when a person quantumly travels through time and space, the soul energies of that individual are changing places with one of his counter-parts in either a past, present, or future parallel universe. For example, if each of these universes is labeled, (X1), (X2), (X3), (Y1), (Y3), (Z1), (Z2), (Z3), and our present universe is labeled (Y2). A person in universe (Y2) can exchange places with one of his counter-parts or doubles in universe (X1), (X2), or (X3), and at the same time one of your doubles which occupies the (X1), (X2), or (X3) universe can exchange places with yourself in the (Y2) universe. See illustration.

So in other words, for every action there must be an equal and opposite reaction, otherwise our universe would collapse. I would also like to point out that this kind of time travel would also explain why people think they have been some place that looks familiar where they never have been before. Maybe perhaps the person in the (Y2) universe is quantumly connecting with one of his future counter-parts in universe (Z1), (Z2), or (Z3). It gives food for thought.

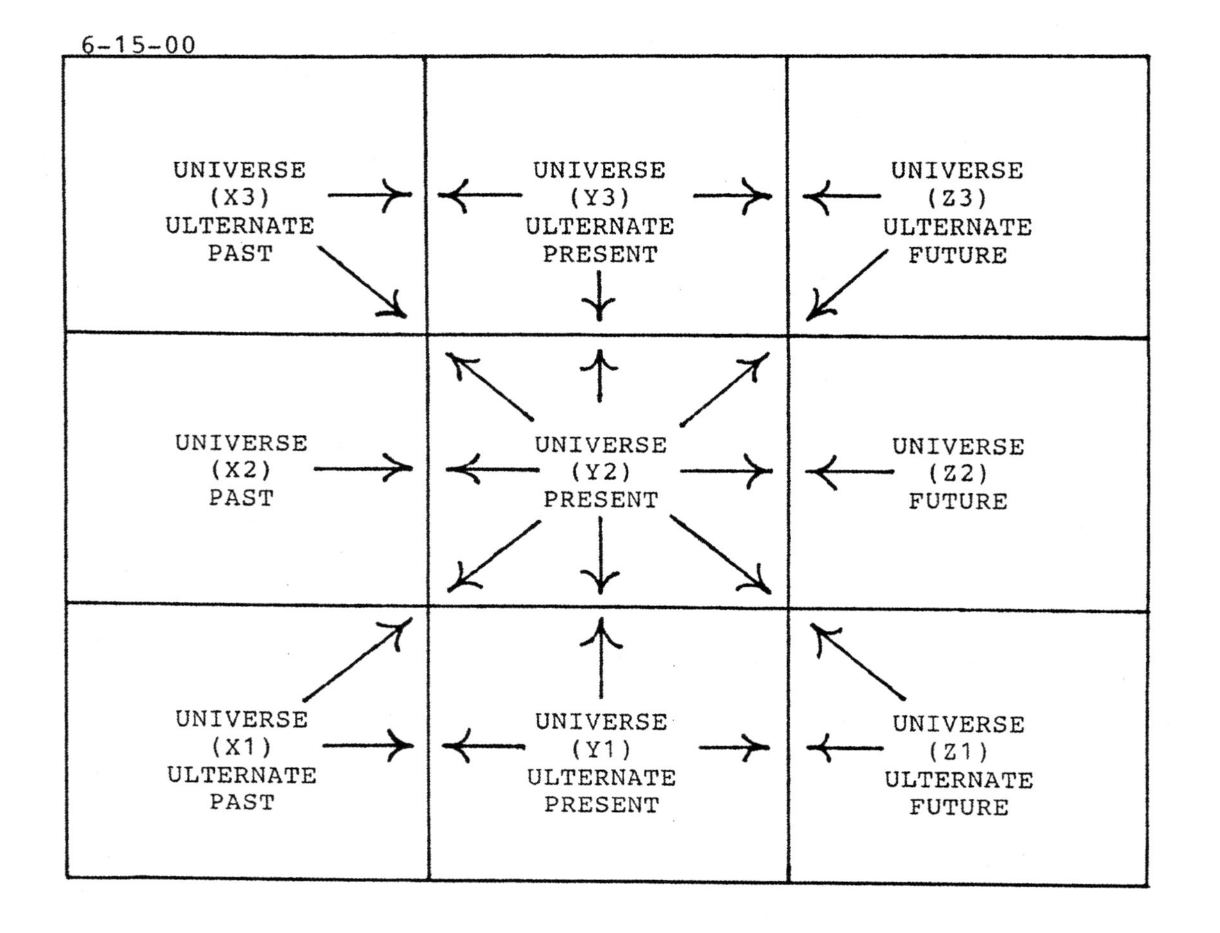
6-15-00
UNIVERSE (X3) ULTERNATE PAST
UNIVERSE (Y3) ULTERNATE PRESENT
UNIVERSE (Z3) ULTERNATE FUTURE
UNIVERSE (X2) PAST
UNIVERSE (Y2) PRESENT
UNIVERSE (Z2) FUTURE
UNIVERSE (X1) ULTERNATE PAST
UNIVERSE (Y1) ULTERNATE PRESENT
UNIVERSE (Z1) ULTERNATE FUTURE

Quantum Time Travel

Now in reference to physical time travel, quantum time travel is similar in many respects except for one difference, one is linear and the other is non-linear. In other words, if a person physically traveled through time you can easily pick up a newspaper to check the date to see how far you have moved through time. However, with quantum time travel you cannot do this. As a matter of fact the exact opposite occurs. For example, let's say you wanted to quantumly travel to the year 1981. As soon as you arrive at the other body, provided that his soul is willing to exchange places with your soul, you would be able to witness the same events you remembered taking place back in the year 1981. But when you go to pick up a newspaper you will find that it is dated 2000 A.D. You then ask yourself the question, why is this so? This is because when you quantumly travel through time you can only move into universes where the events are pushed back but not the date.

Quantum time travel has one distinct advantage, not only is it safer, but you can walk among others without being detected. Not only that, but when you physically travel through time you run the risk of being tracked down by either the aliens or the government. So if you quantumly travel through time you shouldn't have anything to worry about.

N.B.- See next page for a detailed description of how your soul quantumly exchanges places with your counter-parts soul in a parallel universe.

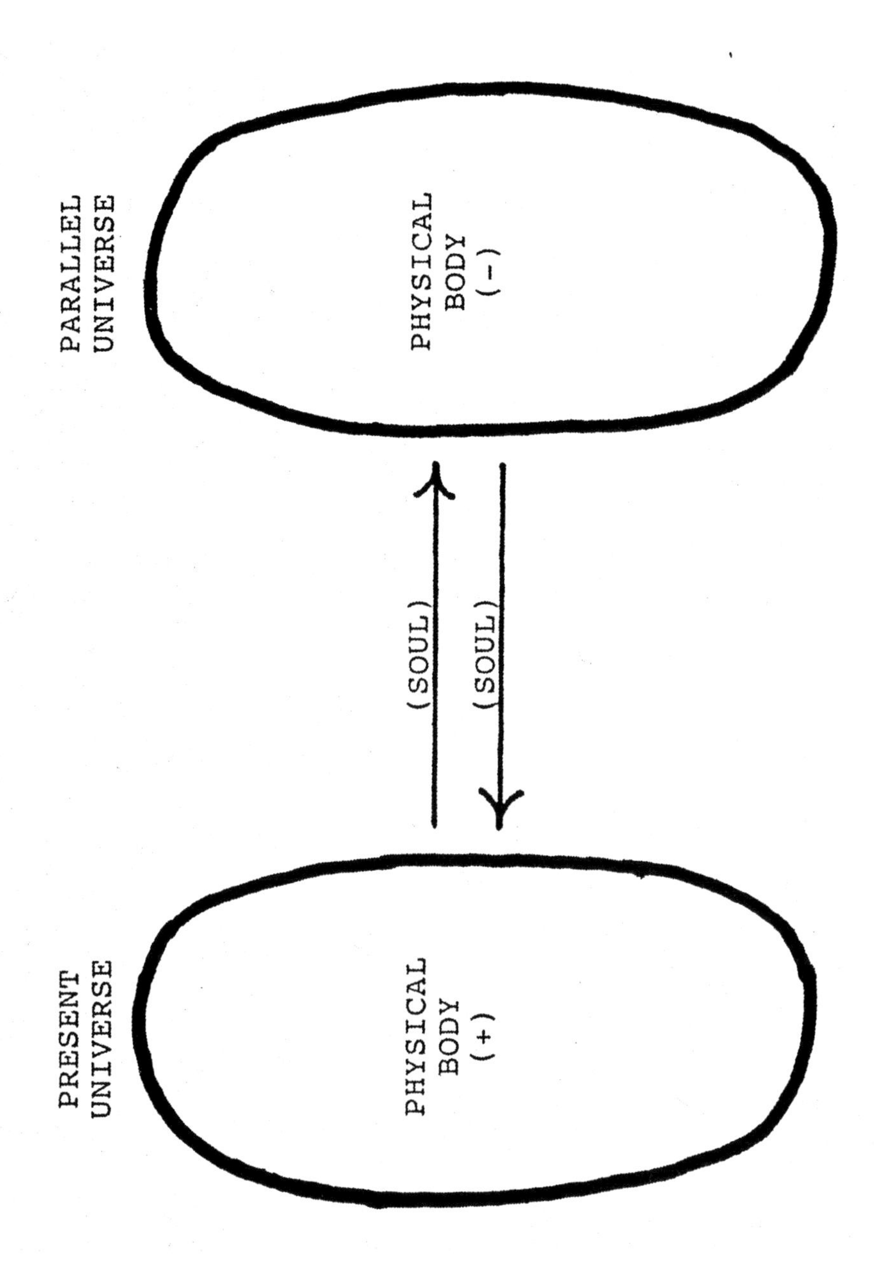
PARALLEL UNIVERSE
PHYSICAL BODY (-)
(SOUL)
(SOUL)
PRESENT UNIVERSE
PHYSICAL BODY (+)

Due to the lack of knowledge on the matter there is also another subject, which needs to be examined, that also deals with quantum time travel. The subject which I refer to is reincarnation. I am not saying that it is a total impossibility for with God all things are possible. However in connection with people who have undergone hypnosis in order to re-call their past lives, I feel that two things occur while they are in that hypnotic state. First of all if the hypnotist does not have a faith in Jesus Christ, then the person who he has hypnotized runs the risk of getting possessed by a demon. And if the hypnotist is an alien god only knows what might happen. Although on the other hand if the hypnotist has a faith in Jesus Christ then the most that could happen is that the person will quantumly connect with one of his past doubles who is currently alive in some parallel universe.

Now what most people don't realize is that most hypnotists are actually aliens in disguise. The reason for this assumption is based upon the fact that some aliens use a form of hypnosis to make people think they are human. In other words they can get into your heads and make you see what they want you to see. And I would not make this claim unless I had first hand experience with these so-called devils. So whatever you do stay away from these hypnotists, because if you don't something bad will happen sooner or later.

How to Convert a Radio into A Time Machine

This device is so simple to build you probably won't believe it. All you need is an ordinary radio where the speaker attachment can be easily removed. Next, rip off the back panel so that all the components are exposed. Walk into your closet and ask Jesus Christ to bless it by praying over it. Then with some salt water sprinkle the components 3 times blessing it in the name of the Father, Son, and the Holy Ghost.

After the unit has been blessed, remove the speaker from either the back or the front side. Afterwards re-connect the back panel to the radio. If the speaker is difficult to remove you may have to use a soldering gun on the (positive) and (negative) terminals to where the speaker connects to the radio. Next, remove the item and re-connect the speaker to both the (positive) and (negative) terminals to where it was connected. When doing this use 2 pieces of insulated copper wiring. Also make certain that the wire being used is at least 7 feet in length. This is so that the speaker can be extended from the radio when connecting it to the stomach chakra or navel area.

Now your all set. All you have left to do is to turn on your radio and radionically tune each of the dials while stroking the upper portion of the radio until you get your stick reaction. When doing this the question should be stated in the following manner. Jesus what are the rates that will teleport my astral body to (year), (month), and (day). While doing this the speaker should remain on your navel area.

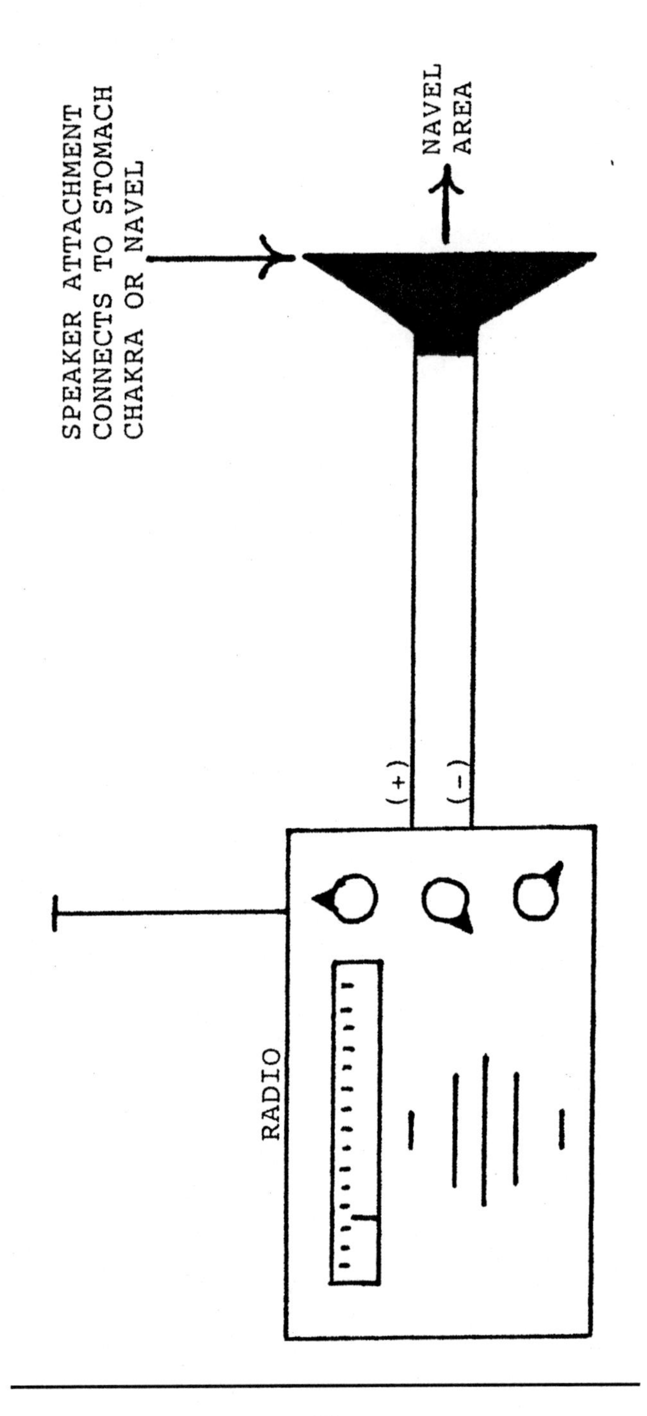
SPEAKER ATTACHMENT
CONNECTS TO STOMACH
CHAKRA OR NAVEL
NAVEL
AREA
(+)
(-)
RADIO

After the rates have been found for an out of the body experience, you then find a comfortable place to relax in order for the energies to take effect. When relaxing try to place yourself in-between the normally awake and normally asleep mode. Once you get there the rest is automatic.

As for physical time travel, all that is needed to be done is to activate the radio over a vortex or grid point on the day of the full moon right at sunrise.

If you decide to pursue the physical aspect of time travel it is *of* the utmost importance that the vortex or grid point should be blessed with salt water. When doing this ask Jesus to bless it by praying over it, then with some salt water sprinkle the area 3 times in the name of the Father, Son, and the Holy ghost. Once the area has been blessed nothing negative will be able to attack you.

Another aspect of traveling through time, especially when the radio is activated over a grid point, is that you may experience a time delay effect. In other words, if the unit lets say is activated for the space of 3 minutes and nothing happens, then you must wait for additional 8 to 10 minutes before you get teleported (I.E. After the unit has been shut down). The reason for this is because it sometimes takes awhile before the doorways will line up. But anyhow once you see the flash of white light you will be instantly teleported through time.

How to Construct the TTC-2000 Time Travel Calculator

This device even though it has not been fully tested as yet, is extremely simple to build. All that is needed for parts is a calculator (which is powered on batteries), some 22 gage speaker wire, a copper plate which is connected to the stomach chakra or navel area, and one double terminated quartz crystal. The materials for making the copper plate are sold in sheets of copper foil that can be obtained from most any hobby store. As for the 22-gage speaker wire, this can be purchased through your local radio shack store.

Construction: to build this device solder one end of the 22 gage speaker wire to either the (+) or (-) terminal where the battery is connected to the calculator. Remember! The battery must remain in the calculator at all times, otherwise if you solder the connections while the battery is removed, you may not be able to insert the battery due to the excess solder or wire. Next, take the excess wire that extends from the calculator, and wrap 7 coils around a double terminated quartz crystal. You now solder the end of the wire, which extends from the crystal to the copper plate. See next page.

NOTE: after the unit has been built, walk into your closet and ask Jesus to bless it by praying over it. Then with some salt water sprinkle the calculator and the crystal 3 times blessing it in the name of the Father, Son, and the Holy Ghost. Afterwards allow the salt water to dry before using. Now you're ready to program the crystal. See next page.

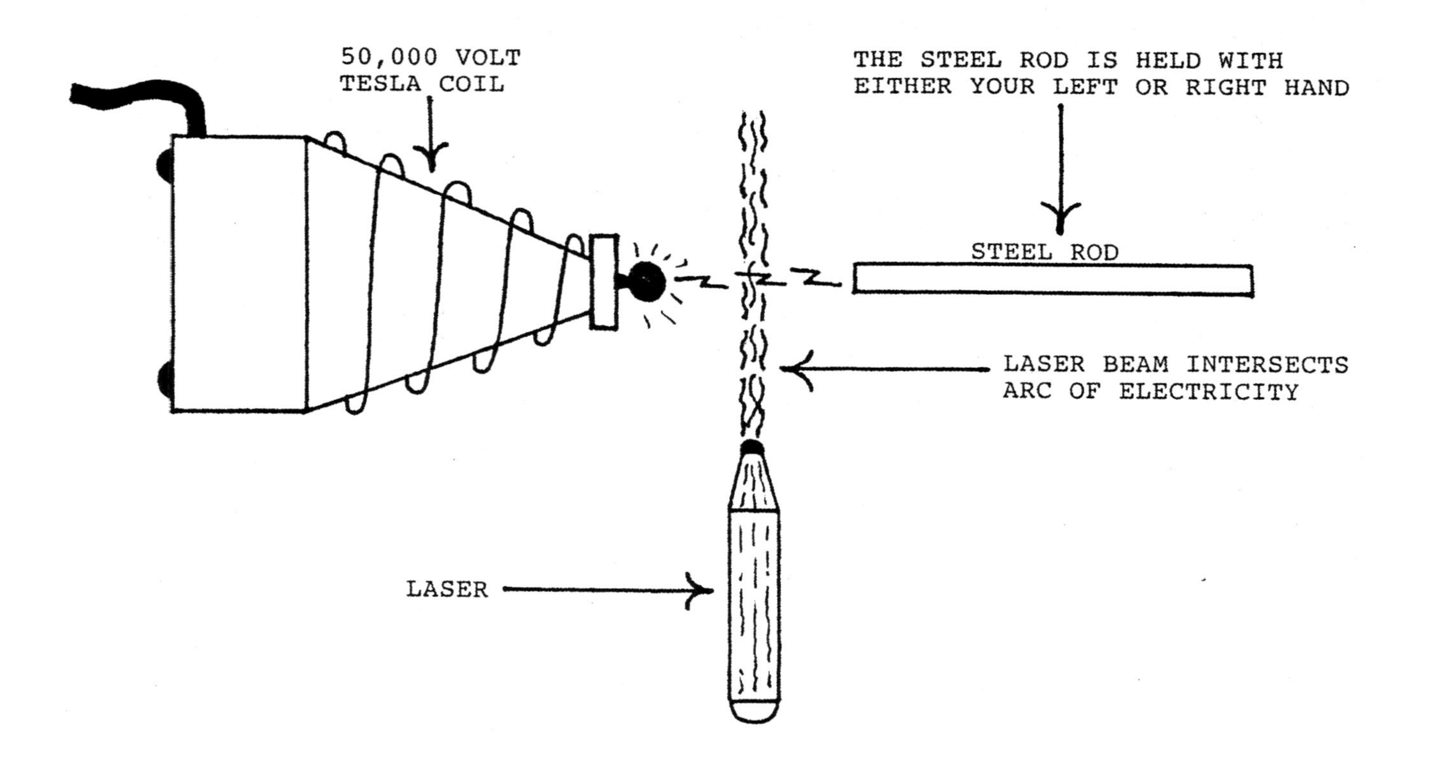
50,000 VOLT
TESLA COIL
THE STEEL ROD IS HELD WITH
EITHER YOUR LEFT OR RIGHT HAND
STEEL ROD
LASER BEAM INTERSECTS
ARC OF ELECTRICITY
LASER

How to Program the Crystal

To program the crystal which connects to the calculator is simplicity in itself. All that you do is while you are holding the crystal in the hand that you do not write with, stroke the upper portion of a wooden desk with your other hand while repeating the following program:

> *Crystal transport my astral body or aura to (date) but only when the numbers (77777777) have been punched out on the key board of the calculator.*

As soon as your fingers stick on the wooden desk, which you were stroking, this means that the crystal has received your program. Now all you have left to do is to connect the copper plate to the navel area and find a comfortable place to relax in order for the projection to take place.

Remember! The 8-digit sequence will be different for each time period you select and for this reason, it may become necessary to keep a notebook, which will store the 8-digit code for each particular date. (See next page).

NOTE: In case you're wondering, the reason for holding the crystal in the hand that you don't write with is because with this hand a North pole field is projected from the palm outwards. So in other words, the North Pole represents the right hand of God or Yang principle, and the South Pole represents the left hand of God or the Yin principle.

Example for Storing 8 Digit Codes

77777777 .... 1/27/57

10000008 .... 2/16/12

77700000 .... 7/17/54

88888888 .... 9/12/48

77001122 .... 5/14/28

70070707 .... 6/12/69

88808080 .... 8/26/54

33003300 .... 7/14/24

11001111 .... 9/24/57

70000007 .... 6/14/23

444400444 .... 1/27/27

70707070 .... 7/24/59

How to Use a Laser for Time Travel

This little piece of information came from a person who lives in Great Falls, Montana. Apparently he succeeded in transporting a small steel ball 20 minutes into the future where it then re-appeared inside the wall of his house. The equipment, which he used to do this, was a Tesla coil and a pin light laser.

Note: in case your wondering, a 50,000 volt Tesla coil may be purchased through Edmund Scientific, and the pin light laser can be purchased through any Radio Shack store.

Well anyhow the method which he used for transporting the steel ball through time was, that while he was holding the steel ball by a string, he held it within close range of a Tesla coil so that the electricity from the ball of the Tesla coil was arcing to the steel ball which was held by a string. Now while this was being done, he then cut the arc of electricity with his pin light laser.

From what he told me, he had to direct the laser beam at the ark of electricity for the space of 15 to 20 minutes before the steel ball disappeared.

After hearing about this I decided to do an experiment myself. However in this experiment the only thing which varied was the use of the steel ball that was replaced by a steel rod which was held in my right hand. So basically while the electricity was arcing to the rod I cut the arc with a laser beam for the space of 3 minutes. See next page. After the Tesla coil and the laser were de-activated I then waited to see what would happen, and sure enough after 15 or 20 minutes had passed I heard a large crack-ling noise and saw a flash of white light.

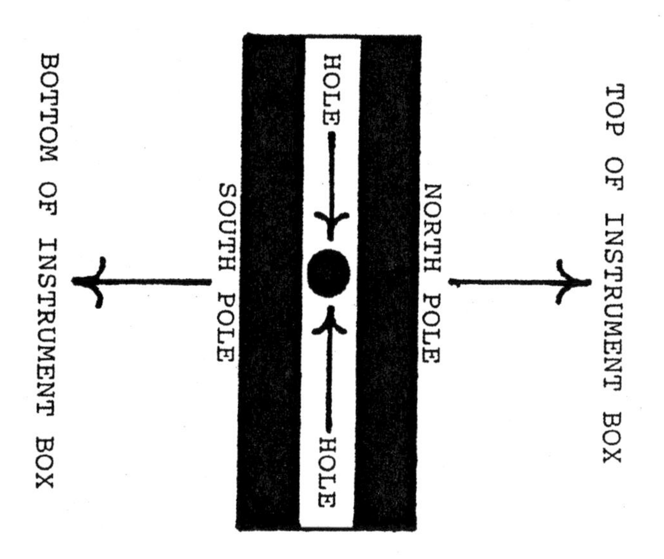
TOP OF INSTRUMENT BOX
NORTH POLE
HOLE
HOLE
SOUTH POLE
BOTTOM OF INSTRUMENT BOX

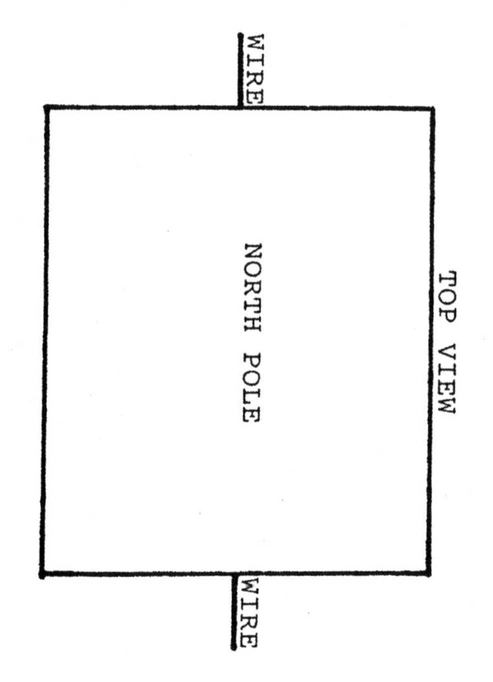
TOP VIEW
WIRE
NORTH POLE
WIRE

Afterwards as far as I could remember the hour hand on my wall clock jumped ahead by a few minutes. But after a few minutes had passed I then returned back to my own present time frame. As far as I was concerned the whole incident was a terrifying experience, one of which I will never forget.

If you haven't guessed it by now, the problem which arises while conducting this experiment, is that you have no way of controlling where you may end up. In other words, it shoots you at random to other time periods or dimensions. By radionically tuning the Tesla coil or the laser by means of a rubbing plate for the year, month, and day you wish to travel to may be the only possible solution for controlling your time frame or dimension. If it doesn't, then the experiment becomes impractical or perhaps even dangerous.

I have never had the opportunity to repeat this experiment a second time. However if I did I suppose I would ask Jesus Christ to bless the Tesla coil and the laser while praying over them, then with some salt water I would bless each of the items in the name of the Father, Son, and Holy ghost.

NOTE - if you fail to do this, you may get trapped in some other dimension or time period. So it is best to tread as though you were walking on scorpions.

How to Construct the BW-2000 Bloch Wall Device

The BW-2000 is fairly simple to build. The only thing which you might have trouble with is constructing the (T1) component. The (T1) component basically consists of a piece of wood which is approximately 1.618 centimeters thick sandwiched between two bar magnets. The piece of wood itself has a hole drilled through the center of it, which is used to house the coil, which is fed through the hole. As for the bar magnets, they should be positioned so that north pole side of one bar magnet is pulled towards the south pole side of the 2nd bar magnet which by their own polarities holds the piece of wood firmly in place. See next page.

Now the reason for using this piece of wood when building the (T1) component is to set up a gap space between the north and south pole side of the bar magnets to create what is known by scientists as the Bloch wall effect. See next page. The Bloch wall or the zero vector is a point which exists between two bar magnets when the north and South Pole sides are facing each other. This is also the place where doorways to other dimensions occur and where all things become infinitely possible. When a Bloch wall is created it usually resembles a figure eight. It is also looks like or resembles a Cadius or morbius field, which may give some added insight as to what is actually occurring.

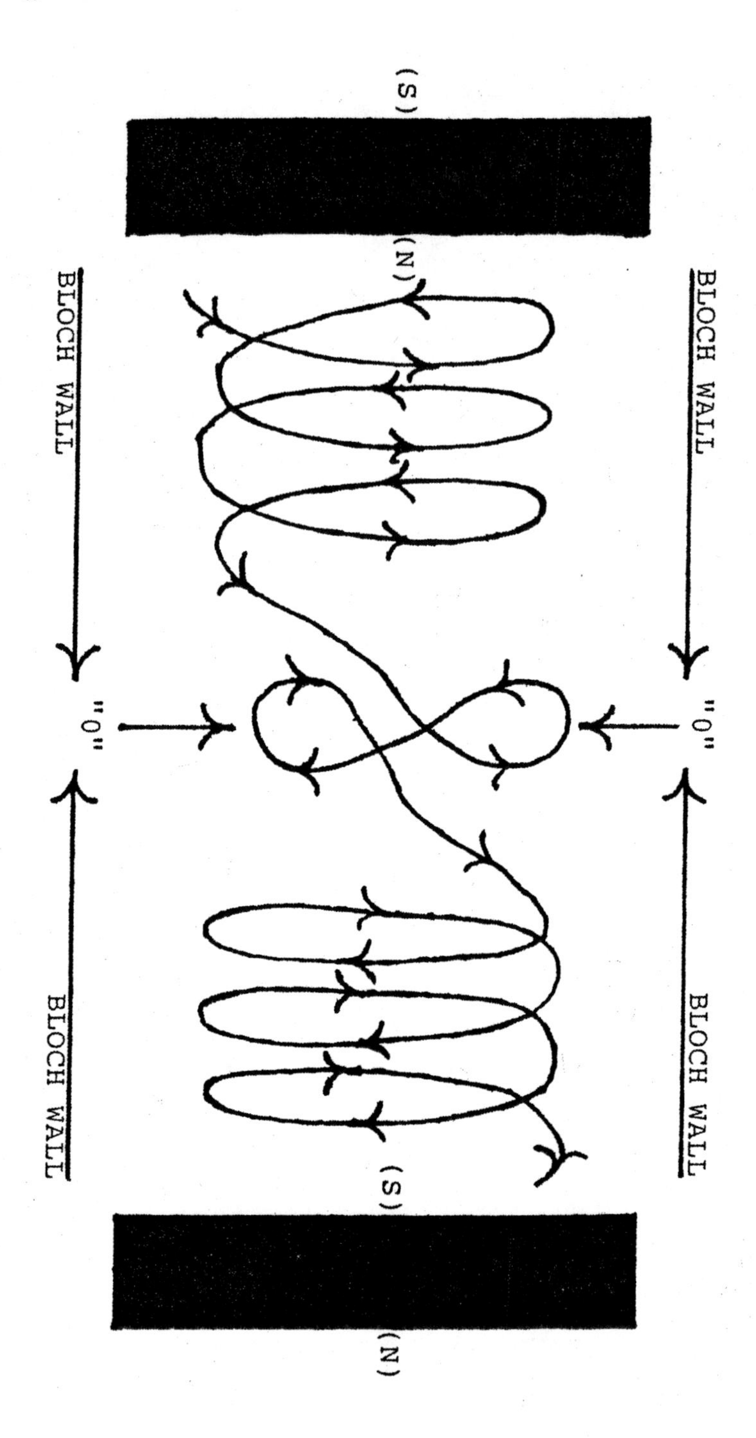
(S)
(N)
BLOCH WALL
BLOCH WALL
"0"
"0"
BLOCH WALL
BLOCH WALL
(S)
(N)

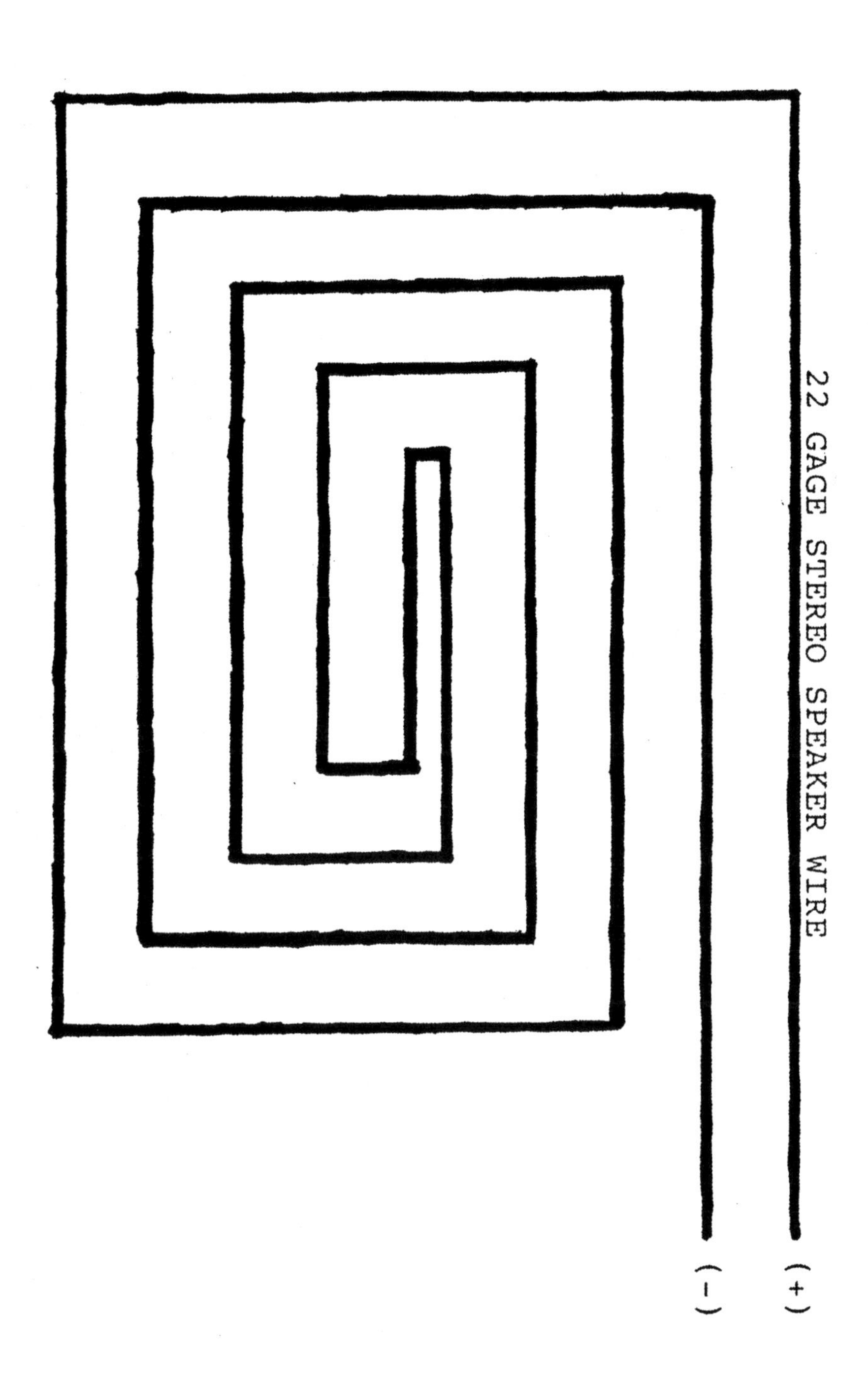
22 GAGE STEREO SPEAKER WIRE
(-)
(+)

The Bloch Wall Device

Be referring to the previous diagram, after the (T1) component is built the south side of the lower bar magnet should be glued to the bottom portion inside the instrument box so that the north pole side of the upper bar magnet is aimed towards the control panel. The reason for doing this is to clear out any negative forces, which may try to interfere when operating the device.

Construction - the parts which are needed to build the Bloch Wall Device are one switch, two 50K pots, one instrument box, 22 gage stereo speaker wire, and on circular copper plate which can be purchased through any Hobby Lobby store.

The bi-filler winding which is made out of the 22 gage speaker wire, see diagram, is positioned underneath the control panel directly below the rubbing plate section. Once the unit has been built it will look something like the drawing titled "The BW-2000".

Operations - after the unit has been blessed and purified with salt water, and also after you have prayed to Jesus Christ, while turning one of the dials stroke the rubbing plates with your fingers while concentrating on the following question: Jesus what are the rates which will teleport my physical body and all of its components to (month), (day), and (year). As soon as your fingers stick on the plate that will indicate the dial setting that will transport you to that date. Then you proceed to the 2nd dial and do the same thing as before.

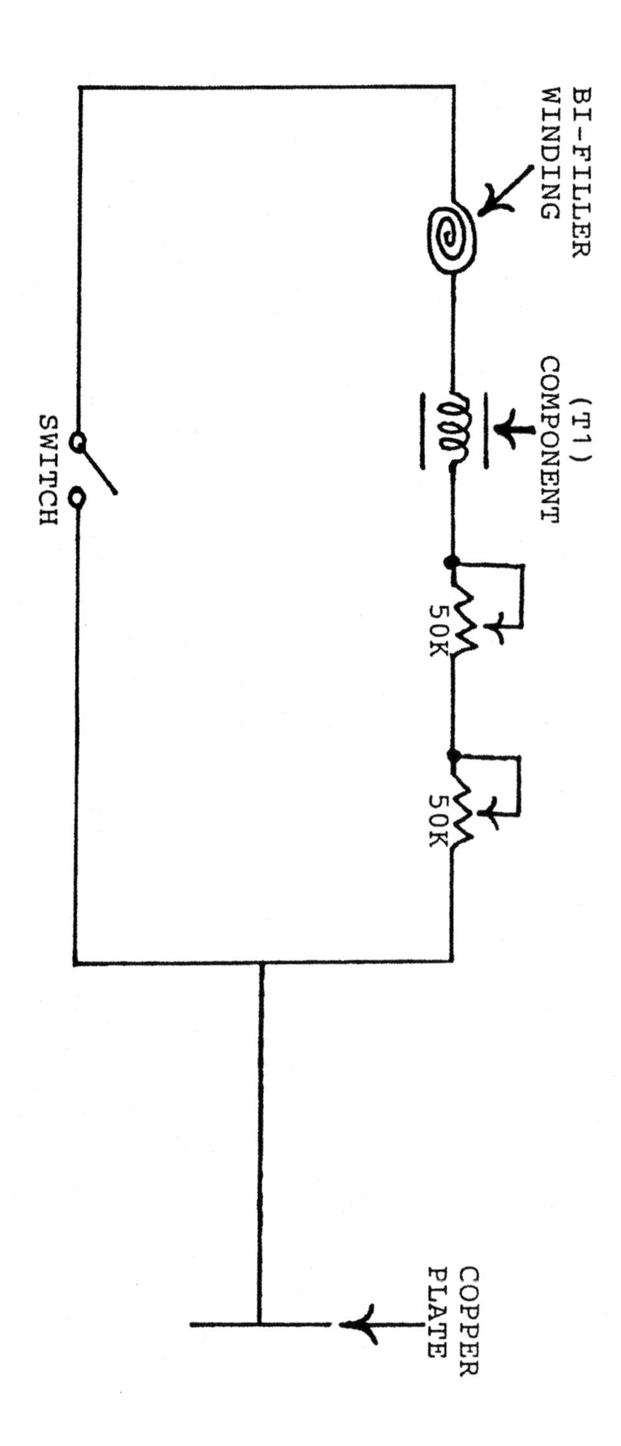
BI-FILLER
WINDING
(T1)
COMPONENT
50K
50K
SWITCH
COPPER
PLATE

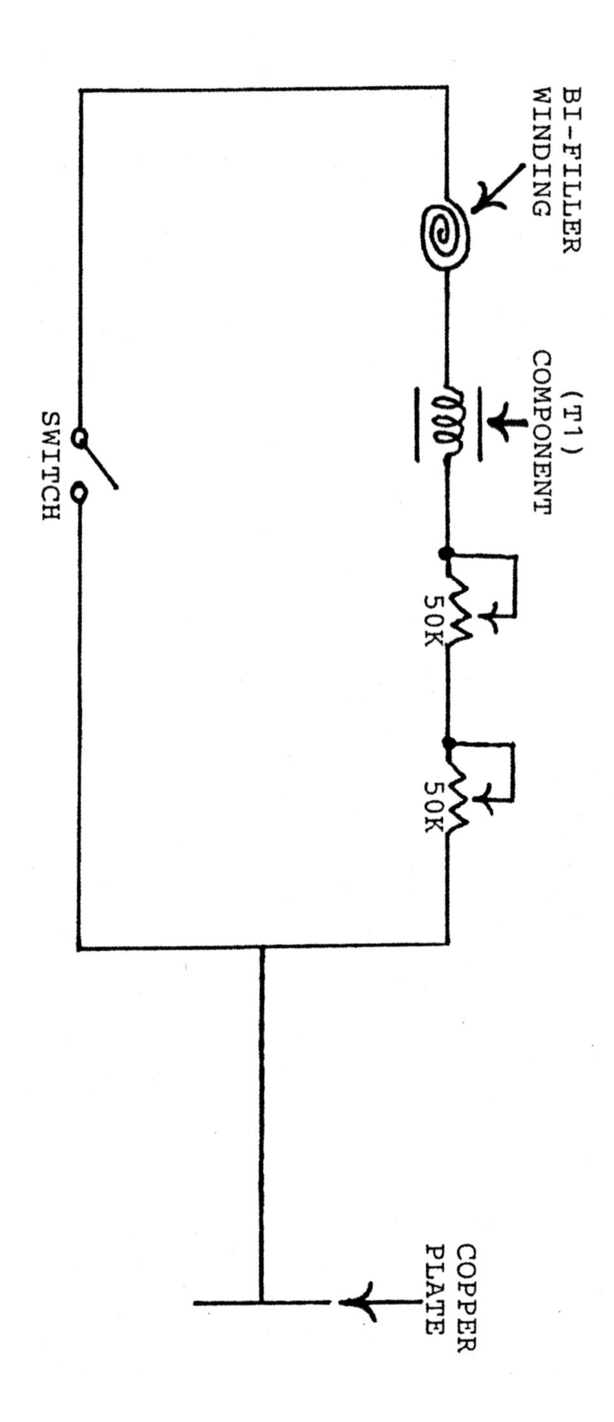
BI-FILLER
WINDING
(T1)
COMPONENT
50K
50K
SWITCH
COPPER
PLATE

After the rates have been found for time travel, the copper plate is not taped over the navel area. If you have done everything correctly it shouldn't be long before you hear a popping noise or see the flash of white light.

It should be added that physical time travel in relation to this device might be dependent on whether or not it is activated over a grid point. If nothing occurs without a vortex then maybe you should try to find one.

Remember! This device is still in the experimental stages so be very careful when using it.

NOTE: In reference to the diagram, when soldering the wires to the 50K pots, they must be connected in the following manner. A failure to connect the pots as indicated below could end up creating some serious problems.

Bottom view of 50K pots

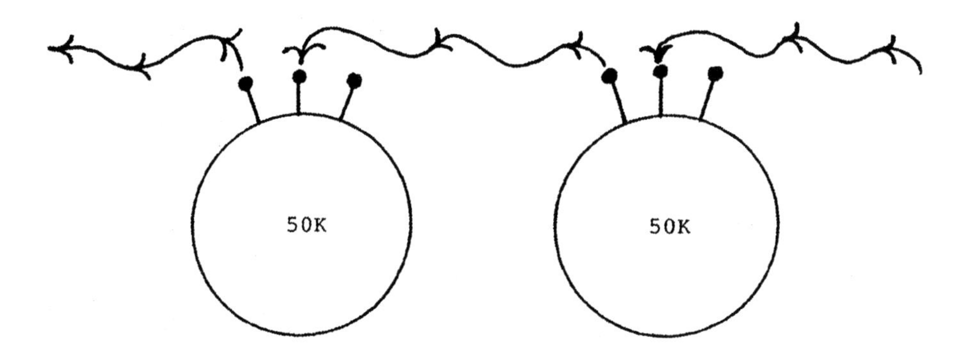

Dear Pat

The Hyper Dimensional Resonator transmits a multi-dimensional Tacyon sine wave which locks onto 20,736,000,000 dimensions, or at least according to my time equations.

Don't forget about __________over in Hawaii who disappeared in front of several eyewitnesses after activating the Hyper Dimensional Resonator over a grid point. Also you can add that part about __________ jump to 2008 A.D., please be careful about using his name.

Another person you could talk about is __________ who also lives in Honolulu, Hawaii. She erased her time line by traveling back 1 year into earth's past and withdrawing $1000 from her own bank account to help out one of her friends who needed some money. After she was transferred to a different past parallel universe, she then returned to her own present parallel universe with no knowledge she even made the jump because in that present parallel universe she never did make jump. However her counter-part from a past parallel universe left some evidence that she did make the jump by leaving a flower chain on a wooden post. She was mysteriously drawn to this place by some unseen force. But what really puzzled her was the fact that the flower chain was composed of her favorite flowers, and on top of that the flowers were still fresh. There was a day or two after she conducted the experiment where she can't remember what happened. There is also some indication that our government may have abducted her. If so they probably erased her memory.

It would also be a good idea not to use her name in the article, as it could get us into a lot of trouble with the military since she used to work for them.

Also don't use __________ name. Otherwise we will be sprouting daisies before the year is over. Be careful what you say.

Sincerely
Steven Gibbs

Biography of Patricia C. Ress

A former journalism scholarship winner and 30-year paranormal investigator, Patricia Griffin Ress met Steve Gibbs in 1991 while writing for and editing *"The Constitutional Liberator."* Steve was one of a long list of research subjects Pat met while serving on the Board of Directors at the Oakcrest Institute in Elkhorn, Nebraska.

She had been in charge of outreach and public relations, which enabled her to be in contact with a large number of people in the UFO/metaphysical community.

Prior to that, Pat had been either a full-time staff member or contract writer for more than a dozen years for several papers and newsmagazines. A member of a prominent medical family, she had also trained and worked as a medical Lab Technician & Pharmacy Technician. In Omaha she worked for 11 years in the travel & transportation industry.

Her first experience in metaphysical writing came when she wrote a column of *"The Strange and Unusual"* for the "Nebraska Voice," a statewide entertainment magazine. She has since written for over 30 regional, national and international publications – most *recently URI Geller's Encounters Magazine of Bournemouth, England.* And since 1998 she has been listed in the Marquis *"Who's Who in America"* as well as *the "Cambridge International Who's Who of Writers and Authors."* A former regional story consultant to "The Mike Jarmus" show (the most widely listened to short-wave radio program in the New York area). Pat is married to Fred Ress. They have 3 children, 5 grandchildren and 3 cats.

Printed in the United States
17949LVS00003B/211-213

9 780759 677074